Joseph Manga Tshomba

Biology and Ecology of the Most Commonly Caught and Sold Fish in Kindu

Joseph Manga Tshomba

Biology and Ecology of the Most Commonly Caught and Sold Fish in Kindu

ScienciaScripts

Cover image: www.ingimage.com

This book is a translation from the original published under ISBN 978-3-639-60833-5.

Publisher:
Sciencia Scripts
is a trademark of
Dodo Books Indian Ocean Ltd. and OmniScriptum S.R.L publishing group

120 High Road, East Finchley, London, N2 9ED, United Kingdom
Str. Armeneasca 28/1, office 1, Chisinau MD-2012, Republic of Moldova, Europe
Managing Directors: Ieva Konstantinova, Victoria Ursu
info@omniscriptum.com

Printed at: see last page
ISBN: 978-620-9-78082-0

Table of Contents

Epigraph

"Whoever shuts their ears to the cry of the poor will cry out themselves and not be answered,"

Proverbs 21:13.

Great battles are meant to be won; sometimes yesterday's failures and today's shortcomings make us fear tomorrow, yet we know full well that victory does not come from what we lack but rather from what we have.

True, we have stumbled, yet we have prevailed. Moreover, courage has shaped us, skill has aided us, and humility has elevated us.

Ir MANGA TSHOMBA Joseph

IN MEMORIAM

To our grandfather Manga Otomba, grandmother Adelaide Namuna, and aunt Sifa Oleka, you should have witnessed the progress we have made from our childhood to the present day, but alas! Providence has taken you from us too soon; may the land of our ancestors once again be a place of peace for you.

Brother MANGA TSHOMBA Joseph

DEDICATION

May all be given to our God for His grace and protection throughout our lives, and may we pray that He may bestow His glory upon us for the remainder of our daily activities.

To our Father, Professor Dr. Manga Okenge Jean Pascal, for all the efforts made and sacrifices offered in the fulfillment of his noble duty and mission as a father, may this work convey our feelings of gratitude as his sons.

To our mother, Mbela Tshomba Lucie, for having cherished us in her womb and for having overcome all the hardships you have endured from our childhood until today, this work is dedicated to you.

To our dearest aunt Emilie Sheda, the second mother in our lives, no one is unaware of your contribution through your love, guidance, and above all, your role in our upbringing. You spared no effort in educating us, and here is the fruit of your labor; please accept our gratitude through these words.

To our dearest and beloved wife, Pipina Eloya Weye Sandra, as well as to my three children, Manga Otomba Johnson, Manga Pipina Joël, and Manga Tshomba Josias, for having supported and consented to this project despite the many hardships and sacrifices required to bring it to completion; may they find in this work an example of courage and perseverance.

ACKNOWLEDGMENTS

At the conclusion of this work, accomplished through enormous sacrifices, we wish to thank Almighty God for protecting us from anything that might have hindered our efforts.

Our most sincere thanks go to the Supervisor of this thesis, Professor Dr. Ir. Nyongombe Utshudienyema Nathan, for agreeing to supervise this thesis; his guidance and scientific rigor were of great importance to us in the completion of this work.

We would also like to thank our Co-supervisor, Professor Dr. Ir. Okitayela Onawoma Freddy, for his guided scientific supervision and, above all, his expertise in the completion of this work;

We further express our gratitude to Professors Baboy Longanza Louis, Kankonda Busanga, Bamawa Maando, Mumba Djamba Antoine, and Batina Agasa, as well as to the Teaching Assistants Solia Edondoto Sylvain and Katusi Roger, and to Dr. Lahay, Brother Ongemba Yalo, and Laurent Mbuyi for the interest they consistently showed in our research. Their field experience and scientific discussions were invaluable to the completion of this work.

We extend our deepest gratitude to the academic and administrative authorities of the University of Kindu in general and to those of the Faculty of Agricultural Sciences in particular.

To our brothers and sisters: Dr. Manga Otomba Jean de Dieu, Dr. Manga Lomame Aimé, Dr. Manga Opondjo Fernand, Prof. Dr. Manga Elongo Jean Pierre, Magistrate Manga Wenye Junior, Dr. Manga Okenge Pascal, Manga Sheda Emilie, Manga Blandine, we thank you for your contributions, in one way or another, to the completion of this magnificent work, which today has become a tangible reality—here is the example to follow.

We would like to extend a special thank you to Maman Atshakawo Tese Charlotte for her kindness, love, and generosity during the writing of this work and throughout our time living by her side. Your advice, your encouragement, and your sacrifices have made us who we are today.

We thank the Kapoli, Ongemba, Pipina, Simene, Kinyonge, Pocho, Katako, Omana, Nyongombe, Hibrahimu, Nyombo, Ngongo, Wembodinga, Foma, Munyantwari, Kamabu, Ekomba, Rachidi, Dudu, Dheda, Mbuyi, Kadiebwe, Sileki, Bwema, Kirongozi, Ekili, Mayani, and Sefora Lembasiya for their support and love;

We would like to thank Mom Riziki Jackie for her dedication and love toward us during this difficult time.

Our thanks go to our cousins for their generosity toward us during this time of hard work.

We would like to thank our friends and comrades in the struggle: Dr. Ir. Onadambo Nyongombe Trésor and Dr. Ir. Ndomba Katokolo Nico.

To all those who contributed in any way to the completion of this work and whose names do not appear here, we extend our sincere thanks.

ABSTRACT

The most commonly caught fish in the central section of the Congo River and sold in the markets of Kindu have been identified. This study made it possible to:

- to inventory the most diverse families in the catches, particularly

notably the Cichlidae with 6 species (17.14%), followed by the Clariidae and Mormyridae families with 4 species (11.43%), then the Clarotoidae and Cyprinidae families with 3 species (8.57%); the families Distichodontidae, Notopteridae, Schilbeidae, Malapteruridae, and Citharinidae are each represented by 2 species (5.71%);

- The remaining families are the least diverse, each represented by a single species, accounting for 2.86% of the total number of fish species inventoried; these are the families Hepsetidae, Anabantidae, Arapaimidae, Mochokidae, and Channidae.

- Regarding physical and chemical parameters:

1. It is observed that the water of the Congo River in this studied section is relatively close to neutrality, ranging from 6.88 to 7.07, a value within the pH range of 6 to 9.

2. The temperature results presented in our study show that the average temperature observed in this section of the Congo River is 26.5°C. The minimum and maximum temperatures range from 23.0 to 29.0°C, respectively.

3. The oxygen results showed that the average oxygen level observed in the river water is approximately 7.4 mg/L. The minimum and maximum oxygen levels are 6.3–8 mg/L, respectively. This well-oxygenated water is conducive to the development, growth, and reproduction of the fish caught there and sold year-round in the markets of the city of Kindu.

Several fishing techniques and gear types were used, including:

- set nets, which accounted for 55.65% of the total 2,318 fish caught;
- gillnets, which captured 26.05% of the fish;
- gillnets, accounting for 13.05%;
- traps, which accounted for 5.13% of the fish.

- The average size at sexual maturity is, respectively: 117.13 mm and 98.72 mm, 54.23 mm and 44.50 mm, 169.98 mm and 152.10 mm, 172.06 mm and 154.0 mm for males and females of *Gnathonemus numenius, Chrysichthys wagenaari, Labeo lineatus, and Clarias gariepinus*;

- The sex ratio in *Gnathonemus numenius* (1:1.8), *Chrysichthys waagenari (1:1.5), Labeo lineatus* (1:1.6), *and Clarias gariepinus* (1:1.17).

- The diet of the economically valuable fish studied in our research is defined as follows: *Clarias gariepinus* is omnivorous with a carnivorous tendency; *Gnathonemus numenius* feeds on insects, leaves, and especially larvae*; the microphagous Labeo lineatus* feeds on dead leaves and insects; *Chrysichthys wagenaari*, on the other hand, feeds on insects, dead leaves, and other unidentifiable items

0. Introduction

The overexploitation of fish stocks in rivers, lakes, and lagoons is now widely recognized at the global, African, and national levels.

This is reflected in particular by a sharp decline in fish catches from one day to the next, despite improvements in the performance of the fishing gear used (Lalèyè, 1995; Levêque and Paugy, 2006). This situation has led to measures for the conservation and rational management of fish stocks and fisheries. These measures rely on precise information regarding the ecology, the composition of the fish fauna, and the level of exploitation in rivers and waterways. Indeed, without a thorough understanding of these parameters, these measures are doomed to fail from the outset.

0.1. Background

Knowledge of African fish fauna has long been of interest to scientists (Gourène *et al.*, 1995), and studies on the biology and understanding of African freshwater and brackish-water fish continue (Crespi, 1998; Lévêque, 1990; Lévêque *et al.*, 1992 in Lalèyè, 1995).

In Benin, several studies have been conducted in recent years on various rivers and bodies of water (Adité, 1990; Niyonkuru, 2001; Ahouansou, 2003; Lalèyè *et al.*, 2004), focusing on major river basins. In contrast, the ichthyological fauna of small water bodies and streams is little known or unknown to the scientific community.

As for the Congo's ichthyological diversity, the fauna remains poorly understood and understudied (Teugels and Guégan, 1994), despite the pioneering work of Boulenger (1901). This is all the more true given that the available information is, in most cases, outdated or fragmentary, or even nonexistent for certain areas.

By way of example, we can cite: Gosse (1963), Poll and Gosse (1963), Matthes (1964), Musala (1989), Nyongombe (1993a; 1994b), Mulangu (2007), Wamuini (2010), Mulangu (2012), and more recently Wembodinga (2014).

The Congo River provides riverside communities with fish as an important source of high-quality protein, and artisanal fishing, fish processing, and trade are undeniable sources of employment and income for these communities. Despite the national and international importance attributed to the Congo River, few studies have been devoted to characterizing the ichthyofauna in Maniema Province, particularly in the city of Kindu and its surroundings.

It is with this in mind that this study, titled *Biology and Ecology of Fish Caught in the Congo River and Sold in Kindu (Maniema/DR Congo)*, is of particular interest. This study aims to inventory the existing fish stocks in this central section of the Congo River and to explore rational management strategies for this fish fauna.

By undertaking this research, we aim to answer the following questions and possibly many others:

1. What are the physicochemical parameters of the Congo River water in the studied section?
2. What fishing gear is used in the studied section?

3. What are the dominant families and species in this part of the Congo Basin?

4. What is the size at sexual maturity of the most commonly caught fish in this environment, and what is their sex ratio?

5. Which fish are sold most frequently, and what are their socioeconomic impacts on households?

0.2. Objectives

For the purposes of our study, we have identified the following objectives:

0.2.1. General Objective

To establish a database on fish from the Congo River, caught and sold in the markets of Kindu, particularly regarding their diversity, the fishing gear used, key biological characteristics of significant specimens, and the physicochemical characterization of the water.

0.2.2. Specific Objectives

The specific objectives of this study are:

- to perform a physicochemical characterization (pH, dissolved oxygen, temperature, transparency, depth) of the river waters in the studied section;
- to identify the fish (families and species) caught using the

most commonly used by fishermen in Kindu;

- to compile a list of the dominant families and species in these catches made by fishermen in order to identify the fish most valued for sale;
- determine the size at sexual maturity, sex ratio, and diet of the most commonly caught fish in the study area.

0.3. Working Hypotheses

We propose the following preliminary responses to the concerns raised by our objectives:

- the ecological conditions would be favorable for supporting this fish fauna in the study area;
- since fishing is a selective activity, the fishing gear (or methods) used would also vary;
- the fish fauna of the Congo River at Kindu would be relatively diverse in terms of families and

species of commercial interest;

- the size at sexual maturity would differ between individuals of opposite sexes, as would the sex- ratio.

0.4. Rationale and Significance of the Study

To feed an ever-growing population such as that in Maniema, fishing is among the recommended means of providing the population with the protein it greatly needs to combat the malnutrition prevalent in rural areas. Furthermore, fishing can enable fishermen to earn money and improve their standard of living.

Very little research has been conducted on this subject in our region, hence the motivation to study fishing, which is the primary activity practiced by the population of Kindu.

This study will contribute to our understanding of the fish caught and sold in this part of the Congo Basin and will also support the sustainable management of existing ichthyological biodiversity, from which certain species can be introduced into fish farming to increase the stock of fish raised in captivity.

0.5. Spatial and Temporal Scope

The sites included in our study are located on the Congo River. A total of thirteen sites were sampled from 2010 to 2014. The selection of these sites was based on the river's hydrography and, above all, the suitability of the sites for adequate fish capture. Information from local fishermen allowed us to prospect these sites in advance to ensure suitable or realistic catch rates. These sites are:

A. Kindu City: 1. Lulindi Beach; 2. Musuku; 3. Mulongoyi; Kunda; 4. Lufubu; 5. Lweki; 6. Ulindi; Lowe (Kibombo); 7. Kindu Central Beach; 8. Mikelenge; 9. Kasuku.

B. Outside Kindu: 10. Mbandaka Beach (5 km); 11. Elila (25 km); 12. Lowa (75 km); 13. Lomami (120 km).

0.6. Structure of the Work

This work is divided into two parts, excluding the introduction, conclusion, and a few suggestions.

In the first part, the literature review is addressed and spans one chapter:

- the first chapter defines some concepts of our study and presents general considerations on water, African freshwater fish, and the sale of fish in Kindu;

The second part consists of original research in which:

- the second chapter describes the research materials and methods;
- the third chapter simultaneously presents, interprets, and discusses the results obtained.

The bibliography and several appendices are also included.

Part 1: Literature Review

Chapter 1: Definition of Key Concepts and General Considerations on African Freshwater Fish

1.1. General considerations regarding African freshwater fish

1.1.1. Fish

1.1.1.1. Definition

A fish is an aquatic vertebrate that is generally oviparous and breathes through gills. It has paired fins—pectoral and pelvic—and unpaired fins—dorsal, caudal, and anal—with skin that is most often covered with scales or covered by a mucous membrane or viscous mucus (Anonymous, 2008).

1.1.1.2. Socio-economic and scientific role of fish

1.1.1.2.1. Socio-economic role

- Fish is a highly digestible food;
- Fish is rich in protein composed of essential and non-essential amino acids capable of combating malnutrition, which has become widespread in our society;
- Fish is a commercial product capable of generating income to improve the producer's social status;
- Fish fat is liquid due to its high content of unsaturated fatty acids, which makes it highly susceptible to oxidation;
- Fish serves as a source of recreation and leisure in sport fishing;
- Fish is the foundation of the canning industry, which generates hard currency.

1.1.1.2.2. Scientific Role

Fish constitute a very important systematic group for conducting research. Due to their varied morphology, physiology, and ecology, they represent a very broad field of scientific research.

1.1.2. Brief overview of the anatomy, physiology, and morphology of fish

1.1.2.1. Anatomy of African Freshwater Fish (Nyongombe, 2009)

Fish found in Africa are classified into two groups: scaleless fish and fish with various types of scales.

1.1.2.1.1. Scales

These are structures that cover the integument of fish and reptiles. In cartilaginous fish (*Chondrichthyans*), they are cartilaginous, and in bony fish (*Osteichthyans*), they are dermal.

The dermo-epidermal scales of *Chondrichthyans* are denticulate cutaneous scales, formerly known as placoid scales. The scales specific to *Osteichthyans* are exclusively dermal; therefore, they lack a denticulate structure and grow continuously (Kestemont, 2009; Harder, 1975).

We distinguish between:

a) cosmoid scales: these consist of four superimposed layers; the outermost layer, which is quite similar to enamel, covers a layer of trabecular dentine (comine), followed by a layer of spongy bone and an inner layer of lamellar bone (isopedine);

b) ganoid scales, characteristic of fossil chondrostomes and both modern and fossil holostomes. They differ from cosmoid scales in that a superficial layer of ganoin (an amorphous, hard, shiny substance) is deposited over the more or less atrophied spongy layers;

c) elasmoid scales, which are very thin and consist only of a highly thinned outer layer of ganoid and an inner layer of lamellar bone. Two varieties of elasmoid scales are recognized depending on whether the posterior field contains spinules—more or less pronounced extensions of ridges formed by the growth zones on the upper surface:

d) cycloid scales, which lack spinules, have a smooth surface, and a regular outline; they are characteristic of lower teleosts (Clupeiformes, Cypriniformes);

e) ctenoid scales, which have spines on the posterior field, known as the acanthogenic field. Ctenoid scales are found primarily in higher teleosts (Perciformes, Scombriformes, Pleuronectiformes).

While ganoid scales are arranged side by side, elasmoid scales overlap on the skin like roof tiles (rhombic or diamond-shaped scales). Figure 1 below illustrates this:

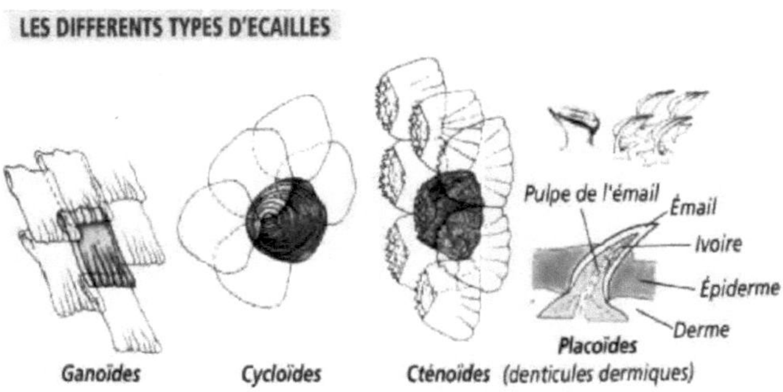

Figure 1: Different types of scales in fish (Source: Anonymous, 2008)

1.1.2.1.2. Fins

These are locomotive appendages that enable swimming and provide body stability. All fins consist of a double skeletal framework: the endoskeleton, made up of cartilaginous or bony elements that support the rays themselves, which in turn support the swim bladder.

We distinguish between the unpaired fins—dorsal, anal, and caudal—and the paired fins—pectoral and pelvic.

a) The unpaired fins: the endoskeleton of the unpaired fins—dorsal and anal—consists of pterygoids of vertebral origin; however, depending on the group of fish, these skeletal elements are more or less connected to the vertebral column.

A special case is the adipose fin, which is located behind the dorsal fin in Salmonidae, Characidae, Siluridae, etc.; this soft fin has no skeletal framework, except for a few actinotrich rays (rays that support the fins of cartilaginous fish and the embryos of bony fish; it is composed of scleroprotein.

The caudal fin is the one that has retained the closest connection to the axial skeleton. This skeleton consists of vertebrae whose neural and hemal arches serve as supports for the pterygoids, which in turn bear the rays.

The caudal fin of most cartilaginous selachian fish and sturgeon is called heterocercal or asymmetrical.

The caudal fin of bony fish is called homocercal. It may be rounded, notched, or truncated.

a) Paired fins:

We distinguish:

- Ichthyopterygia is the paired fin found in most teleost fish; the basal elements are absent, and the pterygiophores attach directly to the girdle. Ichthyopterygia is also known as ichthyopterygia or a rayed fin;

- The archipterygium, also known as the biseriate fin, is characteristic of the Dipneustes; its endoskeleton consists of segments, each bearing two opposite pterygophores, to which the rays are attached (Harder, 1975).

1.1.2.1.3. Lateral line

This is a sensory structure unique to aquatic vertebrates, whose receptors are the neuromasters. These have a cephalic epidermal origin, even though in adults they are located in the trunk or caudal region.

Neuromasters (cutaneous sensory organs of the lateral line) may remain superficial or sink into the bottom of pits or grooves opening outward through two pores.

1.1.2.1.4. Teeth

In fish, teeth lack enamel and are replaced very frequently throughout the fish's life. Teeth may be embedded in the lower or upper pharyngeal bones, and together they form the grinding organ known as the pharyngeal tooth.

1.1.2.1.5. External gills

These are generally temporary aquatic respiratory structures found in the embryos and larval forms of fish and amphibians. There are four pairs of external gills in fish. Branchiospines are protrusions located on the inner, concave edge of the gill arches. They are generally arranged in two rows, anterior and posterior, which interlock with one another, but they may be present on both the posterior and anterior edges.

The gill spines in a single row form a gill comb or rake that partially closes the gill slit and acts as a filter for food particles contained in the water surrounding the gills.The number and shape of the gill rakers are related to diet; indeed, species that feed on large prey lack them, while those that feed on small prey have numerous, small gill rakers that are tightly packed like the teeth of a comb. The position, number, and shape of the gill rakers are very useful distinguishing characteristics for identifying species.

1.1.2.1.6. Gill cover

This is a structure found exclusively in fish and lobe-finned amphibians, which covers the gills, leaving only a posterior opening, or spiracle. In bony fish, it is supported by skeletal elements articulated to the hyomandibular bone and may be extended by the branchiostegal membrane (a structure supported by bony skeletal elements located in the ventral part of the hyoid arch).

1.1.2.1.7. Barbel

These are thread-like organs, generally located around the mouth, and endowed with sensory functions. Each barb may contain a skeletal, fibrous, or cartilaginous axis, muscle masses, and primarily epidermal and dermal tissues containing either taste buds or tactile corpuscles. The number, position, and length of the barbs are variable.

1.1.2.1.8. Gill slit

This is a lateral structure that connects the pharyngeal cavity to the outside and is used either for feeding (filtering microscopic food: microphagy) or for respiration. The gill slits are separated from one another by vertical pillars called gill arches. On these pillars develop the gill structures located behind a more dorsal opening called a spiracle or vent. In bony fish, the gill slit region is covered by the opercular apparatus, which leaves only a pair of posterior openings visible.

1.1.2.1.9. Inner ear

The part of the ear that contains the auditory receptors. In fish, the ear consists solely of the inner ear, also known as the labyrinth.

1.1.2.1.10. Weber's ossicle

Cypriniformes were formerly known as Ostariophysi, from two Greek words meaning "ossicle" and "bladder." This is the double chain of ossicles connecting the swim bladder to the auditory organs of teleost Cypriniformes. It is Weber's ossicles that explain the former name of the Clupeiformes order, "isospondyls," meaning "with identical vertebrae" (Harder, *op. cit.*).

1.1.2.1.11. Physiology of Fish

1.1.3.1. Physiology of respiration

Respiration encompasses all the physicochemical and biological mechanisms that provide living organisms with the energy necessary for their metabolism (as defined by modern physiologists). Biologists generally distinguish between two main types of respiration: aerobic respiration and anaerobic respiration. Respiration refers to the intracellular reactions that produce energy through the oxidation of organic compounds, which release electrons accepted by oxygen. This mode of respiration therefore requires the use of oxygen from the environment and the release of combustion products, primarily carbon dioxide (CO_2), hence the existence of an exchange system known as the respiratory system. Figure 2 below illustrates this:

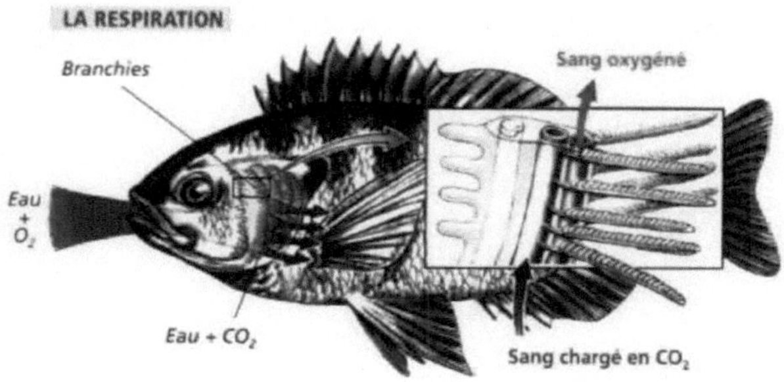

Figure 2: Respiration in fish (Source: Anonymous, 2008)

Anaerobic respiration, in which electrons are captured by nitrates, nitrites, or sulfates, occurs in certain plant species. In animals lacking sufficient oxygen, energy can be derived from lactic fermentation, which is considered a form of anaerobic respiration but is actually complementary to aerobic respiration. Fermentation occurs when there is not enough oxygen available and ceases when oxygen levels become sufficient again; the subsequent metabolism of lactic acid, a product of fermentation, inevitably involves the use of oxygen molecules.

Studying the relationship between oxygen consumption and the oxygen content of the environment allows us to measure a fish's ability to sustain a given metabolic rate. Under hypoxic conditions, certain species have solved the problem of survival in the most unfavorable oxygenation conditions (marshes, drying out, etc.) by developing a mode of aerial respiration.

1.1.3.2. Aerial respiration

In aquatic environments where oxygen is severely lacking, some fish can utilize oxygen from the air. They access this oxygen source in two ways:

- either by developing a surface-breathing behavior that utilizes the best-oxygenated surface layer of the water,
- or through the presence of an aerial respiration system. Morphologically, an organ can only be considered to be involved in aerial respiration if it meets two criteria:
- that gas exchange is possible at the level of the organ;
- there must be a capillary network allowing gas exchange between the blood and the air. The aerial respiration system may result from the adaptation of the oral and pharyngeal cavities, the digestive tract, or more specific organs such as the gas bladder and the lung, depending on their origin relative to the pharynx. The former has a ventral origin and the latter a dorsal origin (Harder, 1975).

Here are the different types of adaptations to aerial respiration found in African species or families with representatives in Africa.

a) Adaptation of the oral or pharyngeal cavity:

- modification of the gills to allow for aerial respiration (Symbranchidae: Mastacembelus);
- pharyngeal air chamber (Channidae);
- specialized structure in the opercular cavity (Labyrinth: Anabantidae; tree-like outgrowth: Clarias).

b) Adaptation of the digestive tract: Cobitidae

Gas bladder:

- Osteoglossidae;
- Notopteridae;
- *Gymnarchus niloticus;*
- *Phractolaemus ansorgri.*

Lungs:

- *Polypterus senegalus;*
- *Calamoichtys calabaricus;*
- *Dipneustes: Protopterus.*

The contribution of aerial respiration to oxygen uptake varies among species under normal water conditions. It increases as the oxygen content of the water decreases and can account for up to 100% of oxygen uptake (Harder, 1975).

1.1.3.3. Physiology of Nutrition

1.1.3.3.1. Feeding habits of fish

When considering their feeding habits, there are three main categories of fish: carnivores, herbivores, and omnivores, which eat both plants and animals. Carnivorous fish eat a wide variety of foods, ranging from large mammals that are attacked and reduced to skeletons, such as the genus Serrasalmus, to tiny plankton feeders with small mouths, such as the genus Nannostomus (Harder 1975).

A fish's feeding habits often provide the key to solving its dietary problems. For example, a fish that spends most of its time near the surface and generally feeds on insects that fall into the water or fly close enough to the water's surface to be caught, or on smaller fish that are typically found in the upper layers of the water. Fish that catch insects by jumping are characterized by a large, upward-facing mouth, while the muscles of the pectoral and caudal fins are well-developed to provide the necessary thrust to propel them out of the water.

On the other hand, those that feed on small fish near the surface have an elongated, streamlined shape. The dorsal fin is located fairly far back on the body and is not very prominent. The head is flattened, and the mouth is large. The teeth are generally small and fine, designed to hold the prey before swallowing it whole. The caudal muscles are well-developed, allowing the fish to propel itself rapidly forward and make a sudden, furious lunge when it spots its prey.

This is the case with Gasteropelecus and Carneglella, which are fish that leap out of the water to catch their prey. Other fish hunt their prey by approaching and lunging at them; examples include the "Panchax" type (*Epiplatys, Aplocheilus,* etc.) and ovoviviparous species. There are fish found in the middle zone. In flowing waters, there are well-formed, streamlined fish that are active swimmers, such as Danios (Brachydanio, Danio). The middle zone of calmer waters is generally inhabited by fish, such as tetras, that tend to travel in schools (*Hemigrammus, Hyphessobrycon*, etc.). There is a group that is better suited to inhabiting reed beds or areas overgrown with weeds.

Species found among reeds are generally highly laterally compressed, which allows them to move easily among the vertical stems. *Pterophyllum scalare* and *Symphysodon* are prime examples. Others like to graze on the leaves of bushy plants, gobbling up the food that accumulates there. They have a long, slender body with comparatively small fins. Pencilfish, *Nannostomus,* and *Poecilobryon* species are good examples of this.

The bottom of the water also shelters fish that obtain their food either by snatching it as it sinks to the bottom or by digging through the substrate in search of live food. These fish have a mouth that hangs downward and is often accompanied by barbels. These whiskers are sensory organs and can be used to locate food. This is the case with catfish such as *Clarias gariepinus, , Barbus sp.*, *Synodontis sp.*, Clarotes, etc. Other fish live near the bottom, not for food but because they find it a preferred shelter. They prefer to stay near rocks and other debris to raise their young. Cichlids (*Oreochromis sp., Tilapia sp., Serranochromis sp.,* etc.) are striking examples of this.

There are also some unusual environments where fish are found, and one of these is the underground cave. Cave fish are sometimes completely blind or may lack functional eyes. These fish possess highly developed taste and touch organs; they even seem to be equipped with a sort of "radar" that prevents them from colliding with obstacles in their path. *Anoptichtys jordani* is one such fish.

Some fish in South America and Africa live in streams that dry up during the dry season. Yet, as if by magic, fry appear in these areas as soon as the water returns, and they grow at an astonishing rate, reaching maturity in a very short time. The eggs had been laid in the muddy bottom shortly before the water bodies dried up. The incubation period for these eggs is long, similar to that of Artemia shrimp eggs. They can withstand drought. The parents, whose lives were coming to an end, would naturally die when the water disappeared. Most of these fish have bright colors. Nothobranchius is just one example among many.

In Africa, it is worth noting the existence of a fish that has another way of surviving dry periods by burying itself in the mud, weaving a cocoon, and blowing a hole with its breath, then falling into a state of lethargy during which vital processes cease almost completely. This tiny spark of life can persist for several years. When the waters return, the fish that have spent the hot months in lethargy must break through their mud shell in order to take a breath of air at the surface as soon as possible. *Protopterus sp.* is the common species in Africa (Kestemont, 2009).

Figure 3 below illustrates some examples of fish mouths:

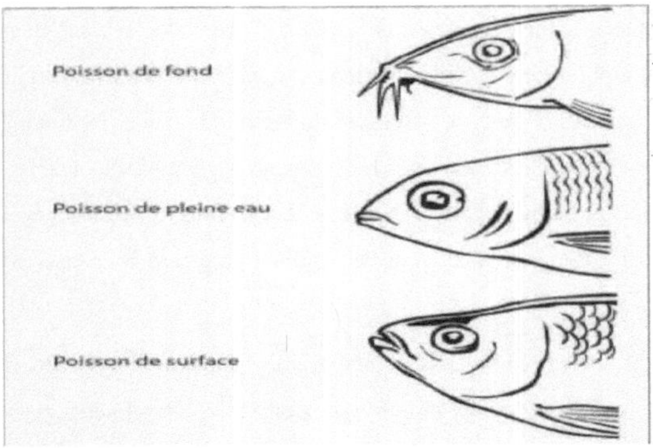

Figure 3: Different types of mouths in fish (Source: Anonymous, 2008)

1.1.3.3.2. The Digestive System

The digestive system comprises the organs responsible for food intake, digestion, and the elimination of feces. It includes the mouth and oral cavity, the esophagus, the stomach, the pyloric caeca, the liver, the pancreas, the intestines, and the colon.

1.1.3.3.2.1. The Oral Cavity, the Mouth, and Associated Structures

a) The oral cavity

The oral cavity is bounded by the mouth in front, the gills on the sides, and the pharynx and esophagus in the rear.

b) The mouth

The orientation of the mouth opening—whether downward or backward—reflects the fish's feeding habits. The mouths of predatory species open forward. Bottom-dwelling fish and scavengers have a mouth that opens downward. Some species, such as carp and sturgeon, have a jointed mouth capable of some extension. This feature allows them to forage and dig at the bottom of the water.

The mouth is bordered by simple folds of skin or fleshy lips (Harder, 1975). The lips enable the grasping and manipulation of food. The position and shape of the lips can serve as an indicator in species differentiation, particularly among sticklebacks (Scott and Crossman, 1974).

The tongue is located on the floor of the oral cavity. It is generally underdeveloped. In conjunction with the movements of the hyoid arches, the tongue helps transport food toward the pharynx.

Some species, generally carnivores, possess teeth that aid in capturing prey. The size of the teeth varies from one species to another, but all are firmly anchored to the mandibular or maxillary bones. When present, the dental pads and pharyngeal teeth are located above the opening of the esophagus. These are flat structures, round, oval, or rectangular in shape, rough to the touch, and continuous with the surrounding skin. These pads serve to grind and shred food before swallowing.

The barbels are innervated structures located around the mouth. They are supported by a cartilaginous base. In the Ichthyidae and Cyprinidae families, the barbels are anchored in the bony structure of the skull (Harder, 1975). The number and position of the barbels vary, making them excellent indicators for identifying species and enabling the localization of food (Hasler, 1957; Toshiaki, 1971).

c) The esophagus

The esophagus is a whitish, tubular muscular organ with a smooth surface. Its primary function is to transport food from the mouth to the stomach. Inside this tube, the folds of the esophageal mucosa are longitudinal and contain numerous mucus-producing cells that aid in swallowing (Ellis *et al.*, 1989).

d) The stomach

The stomach is a muscular organ shaped like an L or an O. It is located in the anterior portion of the abdominal cavity, near the liver. Its surface is smooth, and its color ranges from yellow to white. The vascularization of the parietal surface of the stomach is sometimes visible on the outer curvature. Red and sinuous in appearance, this vascularization should not be confused with certain red nematodes frequently found in the abdomen. The stomach is easily recognizable by its thick wall, which gives it a firmer consistency than that of other gastrointestinal structures.

Upon opening, one can observe the folds of the stomach, which typically originate from the esophagus and extend toward the duodenum. The presence of mucus and fluid within the stomach lumen, along with the absence of food particles, may indicate that the specimen has been fasting for some time. The stomach lumen is usually examined to assess for the presence of parasites.

The size of the stomach varies by species and depending on the amount of food ingested. It is a poorly developed organ in some species, such as trout, or simply nonexistent in others, such as carp. Hydrochloric acid and pepsin, secreted by glands in the stomach lining, initiate digestion by breaking down proteins and carbohydrates present in the bolus (Stoskopf, 1993).

e) The Small Intestine and Pyloric Caeca

These tubular organs serve as sites for the absorption of digested nutrients. They range in color from white to yellowish depending on the species. The parietal surface is smooth. The pyloric caeca are located at the junction of the stomach and the small intestine. They form thin, finger-like structures that are dead-ended (blind-ended), varying in size and number depending on the species. When present, these structures serve to increase the surface area for the absorption of digested food. They are absent in certain species such as pike and barbel (Norman, 1931).

The lining of the small intestine has a velvety texture. When the intestine is wide enough.

The length of the small intestine varies depending on diet. Herbivores have a long, coiled digestive system to increase the surface area for absorption. Conversely, the small intestine of carnivores is rather short and straight. In omnivores, the structure is generally intermediate between that of herbivores and that of carnivores.

The mesentery is the thin connective tissue that connects and supports the intestines. It is anchored to the roof of the abdominal cavity and houses all the vessels that supply the intestines. The mesentery should be examined carefully because it is a common site for the encystment of certain parasites.

f) The Colon and Rectum

The colon and rectum are the final segments of the digestive tract. Together, they form the large intestine, a tubular organ that is wider and has thinner walls than the rest of the digestive tract. Its main function is to reabsorb water and electrolytes from the feces. Feces are excreted outward through the anus. The lumen of the large intestine should be examined, just as with other portions of the digestive tract. (Norman, *op. cit.*).

g) The Pancreas

The pancreas of fish consists of a loose aggregate of cells that do not necessarily form an organ visible to the naked eye. In some species, such as Atlantic salmon, pancreatic tissue is incorporated into the architecture of the liver (hepatopancreas) (Lair *et al.*, 1996). In catfish, it takes the form of a tiny nodule in the mesentery of the spleen and liver region (Stoskopf, 1993). The pancreas is responsible for producing and secreting digestive enzymes into the intestines to aid in the breakdown and absorption of proteins, lipids, and carbohydrates resulting from food digestion.

h) The Liver

The liver is an organ located in the region of the stomach, in front of the pyloric caeca. The liver is more or less lobed, and the lobes vary in size and shape depending on the species.

The surface of the liver is smooth and shiny, and its color is uniform. This uniformity is sometimes interrupted by the presence of bacterial granulomas and parasitic cysts. The color may vary by species, ranging from a brownish red in yellow perch (*Perca flavescens*) to pale pink or nearly white in Atlantic poulamon (Microgadustomcod), a species that accumulates large amounts of lipids. Diet, a period of anorexia, or the ingestion of certain toxins can also cause changes in liver color. The edges of the liver are generally thin and well-defined. Rounded edges or parenchymal patterns that do not follow the regular contour of the liver lobes may be a sign of a pathological condition. The texture of the liver is generally firm, but the consistency may vary depending on the species and the amount of accumulated lipids or due to a pathological condition.

The liver is the site of synthesis for proteins, lipids, and carbohydrates derived from the digestion of food in the intestine. Additionally, the liver serves to detoxify toxins and metabolic byproducts present in the blood. The gallbladder is a small, pouch-like structure filled with a yellow or greenish fluid and attached to the ventral or caudal surface of the liver. It is responsible for producing enzymes essential for fat digestion (Lair et *al., op. cit.*).

i) The swim bladder

The swim bladder is a hollow organ located along the spinal column, beneath the kidney. It plays a role in hydrostatic balance, the production and perception of sounds, and the perception of pressure changes (Hasler, 1957). During embryonic development, the swim bladder forms from a diverticulum of the esophagus. In some species, such as salmonids, this connection between the esophagus and the swim bladder is maintained even into adulthood. Physostomate fish belong to phylogenetically older species, such as trout, and possess a connection—the pneumatic canal—that controls the entry and pressure of air into the swim bladder, thereby increasing or decreasing the fish's buoyancy.

Physoclistic fish are those that lack a canal connecting the digestive system to the swim bladder ( ). A special gland located in the anterior wall of the swim bladder regulates gas pressure, primarily oxygen, in these species (Hasler, 1957).

1.1.3.4. Reproductive habits in fish

These habits vary considerably among different genera and, in some cases, among different species within the same genus. We will discuss these different fish individually, but it may be useful at this point to review certain general characteristics of spawning habits. Based on these habits, fish can be divided into three groups.

The first group of these fish lays its eggs and then leaves them to hatch without tending to them:

1. The eggs are laid and either adhere to plants or other objects, or sink to the bottom. Most *barbs* and *characins* spawn in this manner. Some fish lay light eggs that float on the surface. This is characteristic of *the kissing gourami*, *Helosoma temminckii;*

2. The eggs hang either individually or in groups from plant leaves. Many fish in the group called Panchax spawn in this manner;

3. The eggs are laid in holes and buried.

The eggs are deposited on another animal, which serves as a host until hatching. The best-known freshwater fish that behaves this way is the European bullhead, *Rhodeus sericeus*. Its eggs are laid inside a freshwater mussel, and after hatching, the fry swim out without causing any damage to the mussel.

The second group guards its eggs. This duty is performed by one or the other parent, often by both.

1. The eggs are laid on a solid object, such as a rock or a leaf, where they are then guarded by one or both parents. Most Cichlidae and Nandidae behave in this way;

2. The eggs are laid on the surface of a leaf hanging over the water. Afterward, the parents splash water on them to keep them moist until the fry hatch and dive into the water; one example of a fish that spawns in this unusual way is *Copeina arnoldi*;

3. The eggs are deposited in hiding places and guarded. The only difference from the first method is that the eggs are concealed rather than laid on an open surface. Some Cichlidae and Nandidae prefer to hide their eggs in this manner;

4. The eggs are laid in a hole and guarded. This is the case for Tilapia, as well as for certain American sunfish and other Cichlidae such as Geophagus;

5. The eggs are placed in a nest usually prepared by the male. These nests range from underwater structures built by fish that secrete a sticky renal mucus (Apeltes, Gasterosteus) to floating bubble nests constructed by certain Anabantidae, such as species of the genera Collis, Betta, and Trichogaster.

The third group consists of those in which the eggs remain in direct contact with one of the parents.

1. After being laid and fertilized, the eggs remain suspended at the female's ovipositor. However, they generally do not remain there until hatching, as they are swept away at some point. Species of Cubanichthys and Oryzias are examples of this;

2. The eggs are incubated inside the mouth cavity of one of the parents, who opens its mouth to allow them to enter if danger arises. These fish include Geophagus, Haplochromis, and certain species of Tilapia, Oreochromis, and Sarotherodon;

3. The eggs develop inside the female's body. Once the fry hatch, they emerge as fully formed juveniles capable of swimming on their own. These are, of course, the popular ovoviviparous fish, the Poeciliidae.

In summary, based on their reproductive methods, we classify fish as follows:

1. Fish that scatter their eggs. Examples include: *Anabas testudineus, Aphyocharax rubripinnis, Channa asiatica, Pantodon buchholzi, Neolebia ansorgei, Cyprinus sp*., etc.

2. Fish that suspend their eggs. These include: *Aphyosemion sp., Aplocheilus sp., Epiplatys dageti, E. fasciolatus, E. longiventralis, E. sexfasciatus*, etc.

3. Fish that bury their eggs;

4. Fish that store their eggs: *Rhodeus amarus*.

5. Fish that anchor their eggs: *Hemichromis bimaculatus, Nannochromis nudiceps, Geophagus brasiliensis*.

6. Fish that hide their eggs: *Badis badis* (chameleon fish), Nannacara anomala, etc.

7. Fish that spray their eggs;

8. Fish that dig a hole to place their eggs in: *Copeina guttata, Etroplus maculatus,* etc.

9. Fish that build a nest: *Trichogaster leeri* and *T. pectoralis*.

10. Fish that carry their eggs: *Oryzias javanicus* and *Cubanichthys cubensis*.

11. Fish that practice mouthbrooding: *Chromidotilapia guentheri, Oreochromis mossambica, Betta brederi,* etc.

12. Ovoviviparous fish: *Poecilia branneri, Gambusia affinis, and Gambusia punctata* (Humblet *et al.,* 1974).

1.1.3.5. Fish capture techniques

There are several fish-catching techniques, according to Nyongombe (1993a), who attempted to describe the traditional and modern techniques used in Kisangani and its surroundings. We list a few of them below:

1.1.3.5.1. Traditional techniques

1.1.3.5.1.1. Gathering

This technique involves searching for fish hiding under rocks, pieces of wood, in holes, in riparian vegetation, or any other substrate. Also known as hand-fishing, this technique is performed in shallow areas, on the bank, along watercourses.

1.1.3.5.1.2. Netting

This technique involves damming off a section of a watercourse—especially streams—by constructing a fence made of sticks or tree trunks, which is then filled with grass and mud. The water is then drained from the enclosed area, and the fish are collected. It is tedious and risky work, as sometimes nothing is caught at the end of this hard labor.

1.2.2.3.1.3. Fishing with a funnel trap

The funnel-shaped trap is a basket whose opening is protected by a funnel.

Fishing with funnel-shaped traps involves constructing a frame made of wooden slats in shallow areas (where the water depth does not exceed the average height of a man).

One or more exit paths are left through the grid, each formed by a trap, which may or may not be baited. The opening of each trap faces downstream, or even upstream for unbaited traps. In the latter case, any object attempting to pass through the trap remains there, held in place by the speed of the flowing water. This is a form of passive fishing using traps crafted by expert fishermen.

A fish swimming upstream encounters an obstacle—the grid—and searches for a passage, eventually entering the trap through the funnel-shaped opening. This opening prevents it from escaping. It should be noted here that this method is selective and passive, at least for certain species, as the opening of the screen limits the entry of specimens that are too large to pass through. Furthermore, the captured fish are not necessarily attracted by bait; rather, the installation of the barrier forces the fish to follow the path left by the fisherman.

1.2.2.3.1.4. Fishing with Ichthyotoxic Plants

This technique involves harvesting the leaves, fruits, bark, or roots of ichthyotoxic plants. The harvested part is first pounded until it becomes very fine. It is then diluted on the surface of a watercourse, and as soon as the fish become intoxicated and float to the surface, they are collected from downstream to upstream using nets (Mulangu, 2012).

1.1.3.5.2. Modern Techniques

1.1.3.5.2.1. Spearfishing

The gill net is a net woven from nylon thread. When deployed, it takes on a circular shape. It is weighted by a string of lead weights.

When casting, the fisherman winds the nylon line directly around the handle, lifts it, and throws the net. The net covers a certain area of the water and

gradually closes under the weight of the sinkers, trapping fish or any other organisms caught in that area.

1.1.3.5.2.2. Drift net fishing

These are nets woven from nylon thread. They are rectangular and have *Musanga cecropioid* wood floats on the upper part and terracotta weights on the lower part. These nets range in length from twenty to sixty meters.

There are also set nets. These differ from the previous ones in that they are held in place by the weight of the weights, which reach the bottom of the watercourse. Sometimes the nets are tied to rocks or trees on the banks (NYONGOMBE, 1993a)

1.1.3.5.2.3. Angling

This involves attaching a nylon or transparent line to a hook on a rod of varying length depending on the angler. This nylon line, in addition to the hook, is baited with either an earthworm, a banana, pieces of soap, chikwangue, frogs, snails, or crabs (NYONGOMBE, 1993b; NDJAKI, 2009).

When a fish takes the bait, the float moves, and simultaneously the fisherman pulls the line in with a sharp tug.

Line fishing has several variations. It can be drift fishing or bottom fishing.

- The first involves attaching a nylon line with a hook to a large wooden and grass-covered float; this rig floats.
- As for the second technique, it differs from the first in that the nylon line is anchored nearly to the bottom of the waterway, keeping the rig stationary.

1.1.3.5.2.4. Drifting longline

This technique involves a nylon line containing 100 to 300 floats arranged in a long string. A small line with a baited hook at its end is attached to each float.

The bottom longline does not have a string of floats. However, the nylon line carries 100 to 300 baited hooks and is dragged along the riverbed without any weight or ballast attached to the nylon line.

Handlines, lines, and jigging lines are considered responsible fishing gear, and their selectivity depends on the size of the hooks used. They ensure food security for many communities. However, longlines can have negative impacts on the environment. One of the major issues with longlines is related to *"longlines"* (longlines stretching for several kilometers and set on the surface by large vessels). These pose a danger to seabirds, turtles, and pelagic selachians (*rays* and *sharks*). Another impact of *longline* fishing is a sharp decline in populations of large fish.

a. Bottom longlines consist of several hooks and leaders attached to a main line. Longlines are passive fishing gear.

b. Handlines are hand-held lines with multiple hooks.

c. Jigging lines are handlines designed to target cephalopods: octopuses, cuttlefish, and squid (Seck, 1980)

1.1.3.5.2.5. Fishing with a dragnet

A verveux is a nylon net forming a funnel-shaped pocket. A weight is attached to the top of this pocket. The gear is then submerged to the bottom of the water. The outer part collapses and blocks the passage through which the fish entered (Prado and Dremiere, 1988).

1.1.3.5.2.6. Electric fishing or stun fishing

It is based primarily on the generation of an electric current in the organism through the tail, which stimulates sensory or motor nerve cells. Its application is simple. One simply creates an electric field in the river to be surveyed. The field is bounded by two electrodes—the anode and the cathode—which are connected to a portable transformer rated at 1 to 600 volts and operating at a frequency of 80 Hz. Fish caught in the field are attracted to the anode and then become immobilized; already weakened, they are captured using a landing net (Micha and Ruwet, 1973)

1.1.3.4. Fish Marketing

Almost all of the fish produced in Maniema Province is sold as fresh, dried, or smoked fish. It is currently estimated that nearly 70% of the remaining fishery production is processed through smoking, salting, and chilling.

The fisheries sector policy is an integral part of the government's overall strategy for rural development, which is broadly centered on:

- economic liberalization and privatization;
- administrative decentralization and empowering rural communities in the management of natural resources;
- in the fisheries sector, the fundamental objective is the sustainable management of fishery and aquaculture resources.

It should be noted that this marketing sector is gradually beginning to organize itself, as a distinction can be observed between fishermen, buyers, and retailers.

Thus, fishing remains a subsistence activity for the people of Kindu, who earn only enough to survive. However, if this sector is organized and modernized, fishing can meet local demand for fish and generate foreign exchange by selling the surplus abroad (ANONYMOUS, 2004).

1.2. Population of Kindu

The population of Maniema Province is estimated at 1,400,528 inhabitants, including 438,000 in Kindu, with a population density of 10.59 inhabitants per $^{km^2}$; this is one of the lowest in the DRC. The average annual population growth rate is approximately 2.16%.The gender breakdown of this population shows that women (51.1%) outnumber men (48.9%) (Provincial Inspectorate of Agriculture, Fisheries, and Livestock, *op. cit.*).

1.2.1. Diet of the Kindu Population

Generally, cassava is consumed everywhere in varying quantities, as are bananas and rice. Vegetables consist mainly of cassava leaves. Palm oil is used everywhere for food preparation, along with peanut oil to a lesser extent.

Specifically, the staple food among the Warega is plantain, cassava among the Wazimba and Balubalisés, rice for the Arabized groups, rice and maize for the Wazula, and rice and millet for the Bakusu. Animal-based proteins come mainly from hunting, fishing, and foraging (caterpillars), occasionally from small-scale livestock farming (goats, sheep, pigs, or poultry), and sporadically from large livestock (Provincial Inspectorate of Agriculture, Fisheries, and Livestock, *op. cit.*).

1.2.2. Agriculture, Livestock, Fishing, and Fish Farming

1.2.2.1. Agriculture

Maniema is a province whose agricultural potential is well established, as the soil, hydrology, and climatic diversity offer conditions highly favorable to intensive and diversified agriculture. The major characteristic of agriculture in this province is the predominance of the traditional sector over the modern sector, which has been in constant decline, especially following the Zairianization of the 1970s. This traditional sector is characterized by small farm sizes, the use of rudimentary tools, and reliance on family labor. It is the most significant sector in Maniema in terms of production volume and the rural population employed in it (89.9% of the total population).

- Cassava

It is ubiquitous on most farms and is grown for its tubers and leaves, often in rotation with other crops, notably maize, peanuts, and rice. The Kasongo region remains the largest producer of cassava, accounting for 25% of total production. It is followed by: Pangi, Kailo (including the city of Kindu), Kabambare, Lubutu, Punia, and Kibombo.

- Rice (paddy)

It is a staple food just like cassava. Its cultivation is also widespread throughout Maniema. In terms of cultivated area, the Kasongo territory remains in the lead with 46.4% of production, followed by Pangi and Kailo, then other territories: Lubutu, Kabambare, and Punia.

- Corn

Maniema offers an eco-climatic environment favorable to the development of maize cultivation, which is often practiced in combination with cassava and/or peanuts.

Corn is grown throughout the province, where it is consumed as boiled ears and paste. It is also used with cassava in the production of traditional alcohol. However, it is the Kasongo territory that leads in production with 49.9%, followed respectively by Kabambare, Pangi, Kibombo, Kailo, Lubutu, and Punia.

- Peanuts

Peanuts are also grown throughout Maniema. Peanuts are eaten as a snack, boiled, or roasted.

They are also used to make a paste used as an ingredient in cooking vegetables, meat, and fish. Production statistics place Kasongo in the lead with 50%, followed by Kabambare and Kailo.

- Plantains

Like rice and cassava, plantains are a staple food for a large portion of Maniema's population. They are eaten boiled, roasted, or as lituma (boiled and then pounded). Sometimes plantains are used in the production of local liquor and Chikwange. In terms of production, the Kasongo territory leads with 26%, followed by Pangi, Kailo, Kabambare, and Kibombo.

Food crop production in Maniema is insufficient to meet the demand or dietary needs for an adequate and balanced diet.

- Perennial Crops

As with food crops, the traditional sector accounts for the bulk of perennial crop production, including Robusta coffee, cotton, oil palm, rubber, and lemongrass. The latter two crops, grown in the territories of Kibombo and Lubutu, respectively, are currently being abandoned.

1.2.2.2. Livestock

Livestock farming in Maniema is of the traditional type, focused primarily on small livestock and poultry. There are no other organized livestock operations in this province apart from the small farm in Katako in the Kailo Territory, 7 km from Kindu toward Kibombo. Cattle farming, still in its introductory phase, has also been disrupted by the war.

Maniema's livestock consists mainly of local breeds of goats, sheep, pigs, birds, and small farm animals (chickens, ducks, rabbits, guinea fowl, etc.). The region has seen the introduction of a few purebred breeding stock, notably the "Large White" for pigs, the "Golden Leghorn," and the "Rhode Island Red" for chickens. These purebred breeding stock have now been crossbred with local strains and are completely degenerated. Hence, the introduction of new breeds is necessary to improve carcass quality and productivity.

Compared to the population's needs, the availability of meat from domestic species is estimated at nearly 2 grams per person per day; this indicates a severe shortage of meat products, which is sometimes slightly offset by the supply of natural resources (game and fish).

1.2.2.3. Fishing and Fish Farming

Fishing in Maniema is also artisanal in nature and takes place on the Congo River and its tributaries (Table 2).

Table 2: Fishing sites in Maniema.

No.	Territory	Fishing sites
1	Kabambare	Kabeya and Babuyu
2	Kailo	Pembeliba, Ombela, Lokenya, and Tchumbelume
3	Kibombo	Lake Ndjale, Lweki, and the Lomami
4	Kasongo	Kitete, Nyangwe, and Kilindingulu
5	Pangi	Elila, Tsholobo, and Mandala

Provincial Division of Agriculture (2004).

The equipment and materials used by fishermen include: canoes, nets, hooks, floats (usually in a line), and earthworms used as bait. Fish production is very low in the province, and statistics are difficult to track.

As for fish farming, it is practiced at the household level in a very rudimentary manner through the construction of dam ponds, which require less labor than diversion ponds. However, family-based fish farming constitutes a significant source of animal protein and simultaneously generates some income for many households in the peri-urban areas of Kindu, particularly in the territories of Pangi, Kailo, and Kasongo.

It should be noted that this production is still very low due to a lack of knowledge regarding pond construction standards and management practices.

Photo 5: Poorly maintained ponds in the Kindu area

Part II

Chapter Two: Setting, Materials, and Methods

2.1. Description of the Environment

2.1.1. Location

The city of Kindu is bounded:

- to the north, a vertical line running from the right bank of the Misubu River, passing through the village of Keka downstream of the Congo River, toward the left bank of the Congo River to the north-left of the Kindu Airport road, up to its intersection with the Kindu-Lokando road at the Lwama cemetery;

- to the south, a straight line extending from the source of the Mikonde River to its mouth upstream of the Congo River, running from the right bank to the left bank at the mouth of the Mukolotshi River, passing through the railway line to the intersection with the Kibombo Road;

- to the east, a straight line running from the source of the Mikonde River to its intersection with another straight line extending from the Misubu River bridge;

- to the west, a line extending from the Lokando intersection to the Lwama intersection, passing over the Mikelenge River bridge on the Katakokombe road to the Kibombo intersection.

It is located between 2° 57' south latitude and 25° 52' east longitude, at an elevation of 472 meters and covering an area of 101,295 $^{m^2}$.

The city of Kindu consists of three municipalities, one on the right bank (Alunguli) and two on the left bank: Mikelenge and Kasuku. It covers an area of 101·25$^{9\ km^2}$and has an annual rainfall of 160 mm and an average annual temperature of 25 to 27°C, at an elevation of 497 m, or 25°55' east longitude and 2.57° south latitude (Anonymous, 2004).

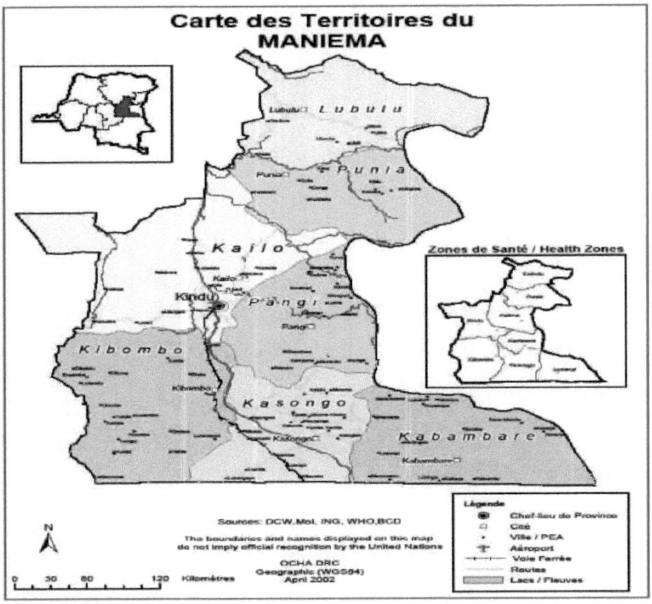

Figure 4: Map of the city of Kindu and the territories of Maniema (Anonymous, 2004).

2.1.2. Topography

The province of Maniema is generally located in the so-called low-altitude zone. Its northwestern part, comprising the territories of Lubutu and Punia, lies entirely within the Central Basin, which has an average elevation of 500 m. Further south, in the territory of Kabambare, the elevation rises gradually to reach 800 m (Anonymous, 2004).

2.1.3. Vegetation

Two major vegetation types cover Maniema: dense rainforest and savanna.

The dense forest covers the Lubutu, Punia, and Pangi territories and parts of Kaïlo and Kibombo. This forest is rich in valuable tree species. Gallery forests cover the Kasongo and Kibombo Territories, as well as part of the Kabambare Territory, where the gallery forest is called "Nyema," from which the name Maniema derives.

Grassy, shrubby, and wooded savannas are found from the south toward the center of Maniema, respectively in the territories of Kibombo, Kasongo, and Kaïlo. In addition to savanna galleries, there are also vast savannas proper, such as the Okona savanna, 132 $^{km^2}$from Kindu, and the Lomami buffer zones in the Kailo Territory.

Grassy, shrubby, and wooded savannas are found from the south to the center of Maniema, respectively in the territories of Kibombo, Kasongo, and Kaïlo.

Photo 1: Savanna forest galleries in the Kibombo Territory and Namoya in the Kabambare Territory

In addition to these savanna galleries, there are also vast savannas proper, such as the one in Okona, 132 km from Kindu, and the buffer zones of Lomami in the Kailo territory.

Photo 2: Savannas (a) of Okona/132 km from Kindu and (b) of Lomani

Photo 3: (a) Dense Bangate forest/Punia territory and (b) old secondary forest of Obokote/Lubutu territory.

Photo 4: Aquatic vegetation found in the study area.

2.1.4. Climate

The climate in the town of Kindu is sub-equatorial, with two distinct seasons:

- A dry season that runs from mid-May to mid-August, accompanied by morning fog and occasional scattered showers;
- A rainy season that occurs twice a year: Season A, which normally begins in mid-September and lasts through January, and Season B, which covers the months of February and March. (Anonymous, 2004).

2.1.5. Hydrography

The Congo River flows through the province from south to north. It is navigable from Kindu to Ubundu (Orientale Province). It is fed by several tributaries, the most important of which are: Lulindi, Musukuyi, Mulongay, Ulindi, Kasuku, Kunda, Lufubu, Lowe, Lueki, and Elila.

It should be noted that the city of Kindu is crossed by the Congo River, which separates the commune of Alunguli from those of Kasuku and Mikelenge. The tributaries of the Congo River located on either side of this section are: the Mikelenge, Lwandoko, Kindu, Misengé, Kange, Mesobo, and Lononga Rivers.

On the right bank (Alunguli commune), the following tributaries are found: the Mesobo, Kabondo, Mangobo, Kamikuga, Kindu, Mesobo, and Mutchondo rivers; while on the left bank, tributaries such as the Mikelenge, Makopo, and Kapondjo are present (Anonymous, 2004).

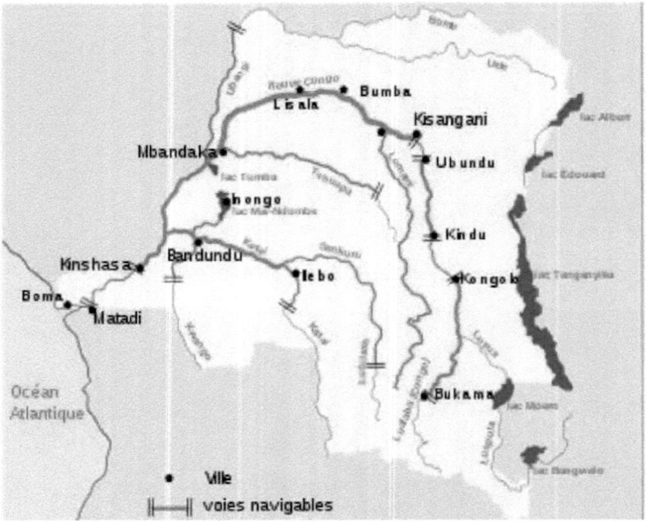

Figure 5: Hydrographic map of the Congo

2.1.6. Soils

Pedogenesis has produced three types of climatic soils in Maniema, namely:

- Arenoferralsol: this group of kaolisaols extends from north to south along the Congo River. It covers eastern Lubutu, central Kaïlo (including Kindu), almost all of Kibombo, and southern Kasongo;
- Ferrisol: it covers part of Maniema, with the exception of southern Kibombo and Kasongo, and northern Lubutu;
- Ferralsol: a group found south of Kabambare. Yangambi-type ferralsol is found north of Lubutu.

The wide range of soil types found in Maniema stems from the variety of bedrock, climate, elevations, and topography. Since soil is an extremely complex system, it can vary considerably even within the same ecological zone, ranging from the sandy soils of the Kibombo territory to the compact clay soils of the large forest areas (Pangi, Kasongo). These clay soils are of excellent agronomic quality, characterized by more lush spontaneous vegetation, easier vegetation growth, and a reserve of mineral compounds, particularly lime. These soils are found in part in Kasongo, Kailo, Pangi, and Kibombo (Anonymous, 2004).

2.1.7. Temperature trends (°C), precipitation (mm), and number of rainy days per month for the past 5 years

Meteorological data are necessary for understanding the hydrology of the watercourse, as shown in Table 1 below

The various measurements taken are necessary for understanding the hydrology of the watercourse or river. Data on precipitation in mm (water depth), air temperature, humidity, and the number of rainy days were obtained from the Kindu/Airport Meteorological Station. An increase in temperature and precipitation and a clear decrease in the number of rainy days are observed.

Table 1: Basic climate values: Temperatures (°C) and Precipitation (mm) for the last 5 years

Month	Average	Maximum	Minimum	Precipitation (mm)	Number of rainy days per month
Jan.	24.6	31.1	21.5	155.1	9
Feb.	24.1	31.5	20.1	161.9	9
March	24.3	31.2	20.6	130.9	9
April	23.8	31.2	20.6	94.5	8
May	23.1	30.8	20.9	138.1	7
June	24.4	31.4	21.6	122.3	7
July	24.1	28.2	20.3	82.3	6
August	23.9	29.4	20.6	110.1	8
Sept.	24.3	30.6	20.1	138.6	9
Oct.	23.8	30.8	21	101.2	9
Nov.	24.1	31.4	21.1	120.2	8
Dec.	24.5	29.8	21.9	113.9	9

Source: Kindu Weather Station/Airport (2014)

2.3.6. Site Selection

The sites covered by our study are located on the Congo River. A total of thirteen sites were sampled from 2010 to 2014.

The selection of these sites was based on the river's hydrography and, above all, the suitability of the sites for adequate fish capture. Information from local fishermen allowed us to prospect these sites in advance to ensure adequate or realistic catch rates. These sites are:

Kindu town: 1. Lulindi Beach; 2. Musuku; 3. Mulongoyi; Kunda; 4. Lufubu; 5. Lweki; 6. Ulindi; Lowe (Kibombo); 7. Kindu Central Beach; 8. Mikelenge; 9. Kasuku.

A. Outside Kindu: 10. Mbandaka Beach (5 km); 11. Elila (25 km); 12. Lowa (75 km); 13. Lomami (120 km).

2.2. Materials, Reagents, and Equipment

2.2.1. Water Analysis

- An oximeter was used to measure dissolved oxygen in the water in mg/L;
- An electronic thermometer was used to measure water temperature;
- An electronic pH meter was used to measure the pH of the water;
- A Secchi disk with a graduated rope was used to measure water transparency;
- A pair of boots;
- Gloves for biosafety.

2.2.2. Catching and weighing fish

- Set net;
- Gill net;
- Gill net;
- Fishing trap.

2.2.3. Identification of fish by family and species

- A digital camera for photographing fish;
- Various identification keys: (Melanie, Stiassny, Teugels, and Hopkins, 2007; Poll, 1957);
- Documentation centers (UNIKIS, UNIKIN, IFA, UNIKI).

2.2.4. Fish measurements

- A caliper with a 15 cm range;
- A 1.5-meter measuring tape;
- A 5 kg scale, accurate to 1 g (Pocket Balance brand);
- A scale with a 50 kg capacity and 10 g accuracy (Pocket Balance brand);
- A 40 cm graduated ruler;
- A measuring tape approximately 1.5 m long;

Photo 7: Taking measurements of a fish

2.2.5. Dissection and identification of stomach contents

- A dissection kit is used to dissect the sampled fish;
- Gloves;
- Scalpels and dissecting knives;
- A pair of scissors;
- Binocular microscope (Eleitz Wetzlar brand, 10x magnification).

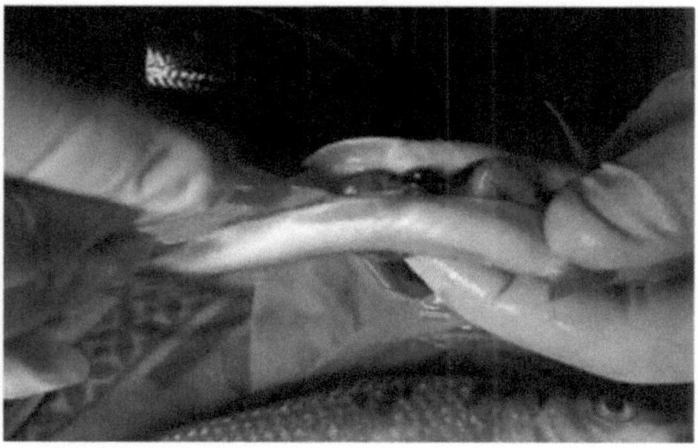

Photo 8: Gutting a fish.

2.2.6. Fish preservation

- 4% formalin was used to preserve the fish specimens;
- Cooler for fish preservation.
- Plastic containers for storing euthanized fish;

2.3. Study Method

2.3.1. Population and Sampling

2,318 fish specimens were collected during our field visits. These fish were classified by family, genus, and species, from which samples were taken. Subsequently, four species were selected for ecological and biological study because they are more commercially viable for potential introduction into fish farming in the Kindu area.

2.3.2. Analysis of physicochemical parameters

These analyses covered temperature, pH, dissolved oxygen, and transparency. The analyses were conducted throughout the year during both the dry and rainy seasons. From January 2, 2010, to February 25, 2014, several field trips were undertaken to assist with data collection.We conducted these analyses at mid-day between 8 a.m. and 10 a.m., between 12 p.m. and 1 p.m., and between 4 p.m. and 5 p.m. on the right bank of the Congo River, in the middle, and on the left bank.

2.3.2.1. Temperature

Temperature has dual significance, both chemical and biological. Chemically, it affects equilibrium constants and reaction kinetics. Biologically, it influences the entire metabolism (embryonic development, growth, respiration, reproduction) and the distribution of animal and plant species in flowing waters.

This explains the interest many researchers (Huet, 1946 and 1970; Decamps, 1967; Hyne, 1970; Lavandier, 1979; Dépasse, 1956) have shown in it. The thermometer was placed in the water at a depth of approximately 20 cm to 1 m.

2.3.2.2. pH

This parameter was determined using a pH meter or even pH indicator paper. The multimeter probe is immersed in water for 1 to 5 minutes, and the reading is taken while the probe is wet; to reset it, the device is then immersed in distilled water.In the case of indicator papers, we immersed them in water for 1 to 10 minutes and compared the results using a color chart.

2.3.2.3. Dissolved oxygen

Dissolved oxygen measures the concentration of dissolved oxygen in water (Rodier, 1984); it plays a role in most chemical and biological processes in aquatic environments. This parameter was measured using a multiparameter probe in water for 1 to 5 minutes and is expressed in mg/L of water.

2.3.2.4. Transparency

In the absence of a 30-cm-diameter black-and-white Secchi disk, we used a 30-cm-diameter plate painted in black-and- , suspended by a graduated string. The plate is submerged or dipped into the water until its visibility limit is reached; at that point, the visibility limit is recorded. This black-and-white painted disc reflects the suspended matter in the water and provides an indication of primary production.

2.3.3. Fishing Techniques

Various fishing techniques were employed: gillnet fishing, trammel net fishing, seine net fishing, and trap fishing. Samples for the ecological study were collected from the different sites, and their results were used for the ichthyological study of the species.

Regarding gillnets and trammel nets, it should be noted that we used nets with the following mesh sizes: 8, 10, 12, 15, 20, and 25 mm. Each net measured 30 m in length and 1.20 m in width.

Fishing traps were also used at Palmier Beach due to the presence of rubble, which facilitates their placement.

2.3.4. Identification of Captured Fish

The captured specimens were identified by referring to the FishBase reference collections and to reference works available in the literature: Boulenger (1909, 1911, 1915, 1916); Lévêque (1990); Poll and Gosse (1995); Teugels and Adriaens (2003*)*; Devos (1995); Mbenga (2004); Tshibwabwa (1997); Melanie, Stiassny, Teugels, and Hopkins (2007).

2.3.3.1. Identification key used to identify fish caught in the central section of the Congo River (Stassny et *al.,* 2007)

1. 5–7 pairs of separate gill slits; cartilaginous skeleton .... **2** Operculum covering

a single gill opening; bony skeleton **4**

2. Gill slits in lateral position; cylindrical body .... **Carcharhinidae**

3. Gill slits in ventral position; body dorsoventrally flattened... **3**

4. Snout extended into a robust, flattened rostrum and armed with strong teeth on each side; tail not clearly distinct from the body **Pristidae**

5. Snout not elongated; head and body forming a rounded-oval disc; tail distinct from the disc, often whip-like **Dasyatidae**

6. Pair of thread-like and elongated, without rays ....

Protopteridae. Paired fins (when present) with rays and fan-shaped **5**

7. Body naked or partially covered with small erectile spines ....**6.** Body appearance varies

entirely or partially covered with scales (sometimes minute) or bony plates or rings

Bony **16**

8. No barbels; body naked or covered with small erectile spines; no adipose fin **7**

Barbels present; body bare; adipose fin present or absent **9**

9. Body very elongated and eel-like .... **Ophichthyidae.** Body of variable proportions but not

eel-like.

10. Pelvic fins absent; body 'spherical'; skin covered with small erectile spines; fused teeth forming a beak **Tetraodontidae.** Pelvic fins

present; elongated body, bare skin; no teeth .... **Kneriidae (including Grasseichthys).**

11. Very long anal fin .... **10.** Short anal fin **11**

10. Very long dorsal fin; adipose fin absent or present, in which case large and radiate; body more or less rounded in cross-section **Clariidae.** Short or absent dorsal fin;

adipose fin absent or present, in which case short and not radiated; body laterally compressed ....**Schilbeidae.**

11. Radiated dorsal fin absent; adipose fin present .... **Malapteruridae.**

and adipose fin present .... **12.**

12. First ray of the dorsal and pectoral fins flexible; three pairs of barbels ....

Amphiliidae. First ray of the dorsal and pectoral fins strong and rigid; two to four pairs of barbels **13**

13. Branched mandibular barbels or mouth modified into a papillary sucker ....

Mochokidae. Unbranched mandibular barbelsnever modified into a sucker **14**

1. Nasal barbels absent; two nares very close together on each side of the head ....

Ariidae. Nasal barbels present or absent; two well-separatedon each side of the head **15.**

2. Nasal barbels present; a single vomerine dental plate ....

Bagridae. Nasal barbels present or absent; vomerine dental plate, if present, composed of two separate dental plates **Claroteidae**

3. Body covered with rhombic scales; dorsal fin with a series of independent finlets, each consisting of a spine followed by one or more soft rays supporting a **membrane. Polypteridae.** Does does not exhibit the combination of the preceding characters

17

17. Elongated body, covered with bony plates in rings, tubular snout .... **Syngnathidae.**

Body entirely or partially covered with cycloid or ctenoid scales (sometimes minute).. **18**

18. Body flattened dorsoventrally; both eyes on the left side of the head .... **19.** Body not dorsoventrally flattened; one eye on each side of the head.. **20**

19. Dorsal and anal fins not fused with the caudal fin; preopercular margin free and visible; anterior end of the lower jaw at the same level as or in front of the upper jaw **Paralichthyidae.** Dorsal, caudal, and anal fins fused; edge of the preopercular margin hidden under the skin; prominent snout.. **Cynoglossidae**

20. Pelvic fins absent ....**21.** Pelvic fins present **22**

21. Eel-like; snout with rostral appendage; elongated anal and dorsal fins fused with the caudal fin; series of independent erectile spines in front of the dorsal fin ....

Mastacembelidae. No rostral appendage; short or absent dorsal fin; long anal fin fused with the caudal fin; no spines in front of the dorsal fin **Notopteridae**

22. Pelvic fins close to the pectoral fins and originating at approximately the same level .... **23.**

Pelvic fins located well behind the pectoral fins.. **39**

23. Dorsal fin without spines..**24.** Dorsal fin with spiny anterior rays (if there are two dorsal fins, the anterior one has spiny rays) **26**

24. Long dorsal fin; anterior pair of tubular nostrils.. **Channi**dae.

Short dorsal fin located toward the rear of the body; non-tubular nares **25**

25. Pelvic fins with long filamentous rays; large pectoral fins .... **Pantodontidae.**

Pelvic fins without long filamentous rays; pectoral fins not enlarged... **Poeciliidae (partly Procatopus).**

26. .First ray of the anal fin simple and flexible .... **27.** Anal fin with one or more spiny **29**

27. Pelvic fins more or less fused, often forming a disc; base of the second dorsal fin longer than the distance between the tip of the fin and the origin of the caudal fin .... **Gobiidae.**

Well-separated pelvic fins that never form a disc; base of the second dorsal fin shorter than the distance between the tip of the fin and the origin of the caudal fin..**28**

28. A single long dorsal fin; dorsal, caudal, and anal fins fused.. **Microdesmidae.**

Two dorsal fins; dorsal, caudal, and anal fins separate... **Eleotridae.**

29. Two well-separated dorsal fins .... **30.** Single dorsal fin or two contiguous dorsal fins

; anterior dorsal fin sometimes reduced to a series of spines **31**

30. Rays of the lower portion of the free pectoral fins.. **Polynemidae.** Pectoral fins

pectoral fins without free rays (Fig. 31) .... **Mugilidae**

31. Body very high and laterally compressed; pelvic fins rudimentary .... **Monodactylidae.**

Does not exhibit the combination of the preceding characters **32**

32. Anal fin with three spines, the first two well separated from the rest of the fin; deeply forked caudal fin **Carangidae.** Continuous anal fin; caudal fin of

various shapes but never deeply forked **33**

33. Two strong spines on the anal fin; rhomboid tail ... **Sciaenidae.** Three or more strong spines on the anal fin; tail

more or less rounded or notched **34**

34. Dorsal fin deeply notched between the spiny and soft sections; a few strong denticles above the base of the pectoral fins **Latidae.** Dorsal fin more or less

unnotched or slightly notched without denticles **35**

35. .A single nostril on each side of the head .... **Cichlidae.** Two nostrils on each side of the head **36**

36. Fewer than four anal spines; scaly axillary process at the base of the pelvic fins .... **37.**

Four or more anal spines; no scaly axillary process at the base of the pelvic fins **38**

37. Mental pores present; no teeth on the vomer; jaws without developed canines ....

Haemulidae. No menton pores; teeth on the vomer; jaws often with well-developed canines **Lutjanidae.**

38. No lateral line; large mouth................................. **Polycentridae.** Two incomplete lateral lines; small

mouth **Anabantidae**

39. Adipose fin present (if exceptionally absent, scales always ctenoid(1) .... **40.**

Adipose fin absent **42**

40. Ctenoid scales; at least a few bicuspid teeth on each jaw; lateral line in the middle of the flanks **Distichodontidae.** Cycloid scales; conical or multicuspid teeth, generally

well developed; lateral line on the lower part of the flanks **41**

41. Dorsal fin inserted distinctly behind the origin of the pelvic fins; a triangular lateral dermal fold on each lower half of the jaw and a smaller one at the tip of the upper jaw; elongated jaws with conical teeth **Hepseticae.** Origin of the dorsal fin

close to the origin of the pelvic fins; no dermal folds on the jaws; short jaws with conical or multi-cuspid teeth **Alestidae**

42. Caudal fin forked or V-shaped ...**43.** Caudal fin neither forked nor V-shaped, sometimes lyre-shaped **48**

43. Ventral part of the body covered with bony scales that may be minute and visible only after staining .... **44.** Ventral part of the body never keeled and lacking bony scales .... **45.**

44. Numerous small denticles on the bony surface of the head .... **Denticipitidae.**

No denticles on the head .... **Clupeidae**

45. No teeth on the jaws, sometimes replaced by keratinized ridges .... **46.** Teeth on the

jaws **47**

46. Cylindrical body with flattened ventral surface; lower mouth with keratinized ridges; numerous small scales; large paddle-shaped pectoral fins in a ventral position.. **Kneriidae (in**

Parakneria). Does not exhibit the combination of the preceding characters .... **Cyprinidae**

47. Mouth with a wide opening; short, radiated dorsal fin, located at the midpoint of the body, well in front of the anal fin; scaly axillary process at the base of the pectoral and pelvic fins .... **Elopidae**

Small mouth; long or short rayed dorsal fin, but always located in the posterior half of the body and of the anal fin; no axillary process **Mormyridae**

48. Elongated dorsal and anal fins in a posterior position; large cycloid bony scales on the body

; lateral line present .... **Arapamidae**

Does not exhibit the combination of the preceding characters; often small and colorful **49**

49. Upper pore of the preopercular lateral sensory system very close to the eye, pectoral fins in a high position **Poeciliidae.** Upper pore not close to the eye, pectoral fins in a

low on the body ....**Aplocheilidae**

1.3.1. Family Arapamidae (Osteoglocidae), *Heterotis niloticus* (Cuvier, 1829)

Description: Body rather elongated, height 3.5–5.0 times the standard length (SL). Head fairly short, 3.5–5.0 times the standard length (SL). Dorsal fin with 32–37 rays, anal fin with 34–39 rays. Origin of the dorsal fin slightly behind that of the anal fin. Scales more or less rough, formula: 2.5/34–40/6. Dermal bones of the head , deeply sculpted. Fine, closely spaced gill spines, 33 (juveniles)–98 on the lower gill arch and 21 (juveniles)–76 on the upper gill arch.

These numbers increase with size. Maximum length: 1000 mm SL. A standard length of 980 mm was recorded for a specimen weighing 10.15 kg. Coloration: uniformly grayish or bronze, with yellowish lips and more or less golden eyes. According to some authors (Blache, 1964; Daget, 1954; Moreau, 1982), this coloration may turn blackish at sexual maturity in juveniles.

3.2.1.1. Family Mormyridae

Bigorne (2003) states that the Mormyridae are fish found exclusively in African freshwater habitats. They are characterized by a reduced posterior body section and associated vertical fins, palatal and lingual teeth that aid in prey capture, and auditory otophyseal connections to the swim bladder. It is noted here, however, that the swim bladder has sent out two diverticula that, during embryogenesis, form two vesicles housed in lateral cranial foramina, thus establishing close connections with the inner ear (the vesicles being located between the semicircular canals).

A paired electric organ of muscular origin is present on either side of the caudal peduncle (Bigorne, 2003). This organ emits weak currents that produce an electric field around the fish. Deviations in this electric field upon contact with obstacles or other fish provide information to the emitting fish via sensory receptor organs located in the skin: the lormyromasts, organs independent of the lateral line. The skin itself has a remarkably thick epidermis that houses these organs. Unlike other electric fish, the power of the Mormyridae's organs is very low (Hopkins et *al.,* 1998).

The body is elongated and comes in various shapes. The head is bare and covered with pores. The entire body is covered with small cycloid scales. These fish have a small mouth that comes in various shapes. The eyes are small with free rims. The fins consist solely of soft rays. A lateral line is present (Nelson, 2006).

According to Gosse (1963), fish in this family are sensitive to dissolved oxygen deficiency. Several authors believe they can be used as indicators of oxygen deficiency in fish ponds (Kandem et *al.,* 1997).

In Central Africa, several genera of fish in the Mormyridae family remain to be discovered (Hopkins et *al.,* 2007); Taverne (1971a, 1971b, 1972); Jégu et *al.,* (1984); Bigorne (1987, 1989, 1990); Boden et *al.,* (1997) and Harder (2000).

According to Kramer et *al.,* (2004); Nelson (2006), the family comprises 18 genera and 201 species. In this family,

Key to the genera

1-A long, tubular, downward-pointing snout; the nostrils are far from the mouth*Campylomormyrus*

-Nostrils are side by side 2

2-Short, stout body; the first two rays of the fin are unsegmented; the nostrils are side by side, with an inferior mouth located below the eye .... *Petrocephalus*

-Terminal or inferior mouth, located in front of the eye; high or elongated body; generally a single, unsegmented ray at the base of the dorsal fin; nostrils separated from each other and from the eye 3

3- Elongated body, 34 to 49 rays in the anal fin and 20 to 35 in the dorsal fin; the snout is not elongated into a proboscis; the anal fin begins well in front of the origin of the dorsal fin, *Mormyrops.*

Campylomormyrus Bleeker, 1874

The head of fish species in this genus is oriented downward. The snout is prominent and tubular. The mouth is lined with a few conical teeth. The dorsal and anal fins originate at nearly the same level and are almost the same size. The dorsal fin is slightly shorter (Hopkins et *al.,* 2007). The chin appendage is covered with electroreceptors used for passive and active electrolocation, likely serving to locate prey buried in the sediments. One species was captured in our study.

Species key

Snout length 148–179% of the length of the post-ocular portion of the

head *C. rhynchophorus*

***Campylomormyrus rhynchophorus* (Boulenger, 1898)**

Observed characteristics

This species has an elongated snout forming a downward-pointing, tube-shaped proboscis. The mouth contains 7 to 10 conical teeth (4 to 6 teeth on each jaw). The fish's body has 26 to 29 rays in the dorsal fin, and 32 to 34 rays in the anal fin. There are 78 to 85 scales along the lateral line and 14 to 16 scales around the caudal peduncle. The specimen exhibits an almost silvery coloration in live specimens and a brownish coloration in formalin-preserved specimens.

Description

Fish belonging to this species have a head length ranging from 32–40% of total body length (TL) and a body height of 22–26%. The eye has a diameter of 10–12% of the head length. The snout is extended into a compressed tubular proboscis, with a length ranging from 150 to 180% of the length of the post-ocular portion of the head, and its smallest diameter is 18–20% of its length.

The small mouth contains conical teeth, 3–5 in the upper jaw and 4–6 in the lower jaw. The caudal peduncle is 350–380% longer than it is high. The longitudinal line comprises 83–90 scales, 22–24 scales between the origin of the dorsal fin and the lateral line, 21–23 between the origin of the dorsal fin and the anal fin, and 16–18 around the caudal peduncle. The dorsal fin consists of 28–30 rays, and the anal fin comprises 30–32 rays. The caudal fin has pointed lobes. They have 46–47 vertebrae (Poll et *al.,* 1982).

Geographic range

This species is also known from the Congo Basin (Gosse, 1984a).

***Mormyrops* Müller, 1843**

According to Bigorne (2003), fish of this genus have a laterally compressed and elongated body. Their body depth ranges from 8 to 21% of total body length. The mouth is prominent and terminal, with 10 to 36 bicuspid teeth arranged in a single row on each jaw.There are 71 to 113 scales. The anal fin is fairly long, with 38 to 51 rays. The dorsal fin is short and begins further back than the anal fin. Two species were collected in this study.

Species Key

1 .-8 scales around the caudal peduncle, 17 to 19 rays in the dorsal fin *M. angolensis*

-11 to 13 scales around the caudal peduncle..2

2 .-22 to 25 rays in the dorsal fin; 36 to 42 rays in the anal fin *M.angullo ide*

***Mormyrops angolensis* Linnaeus, 1758**

Observed characteristics

This species has a height ranging from 16 to 23% of total body length (TBL). The head is elongated, with a concave upper profile, smooth, and indented forward; its length measures 24–36% of TBL and its width 7–16% of TBL. The snout is broad, rounded, and ranges from 21 to 30% of the head length. It contains 21–25 small teeth in the upper jaw and 24–31 in the lower jaw. The dorsal fin is shorter than the anal fin and contains 23–28 rays. Live specimens were grayish in color.

Description

This is a species with a relatively elongated body; its height ranges from 17 to 23% of total body length (TSL). The head is elongated and smooth; its length measures 27–34% of TSL and its width 8–17% of TSL. The snout is broad, rounded, and ranges from 22 to 29% of head length. It contains 20–24 small teeth on the upper jaw and 25–32 on the lower jaw. The eyes are tiny; their diameter ranges from 7 to 13% of head length. There are approximately 6 gill rakers on the underside of the first gill arch. The scales are small, with 90–95 along the lateral line, 14–16/16–20 in a transverse row, and 16–18 around the caudal peduncle. The caudal peduncle is 17–21 times longer than it is high. The dorsal fin is shorter than the anal fin and has 26–28 rays; the anal fin has 45–51 rays. The caudal fin is relatively small.

Geographic Range

The species has a wide distribution; it is found in West Africa and the Congo Basin.

***Petrocephalus marcusen*, 1854**

Members of this genus have a small, inferior mouth that opens vertically beneath the center of the eye or slightly in front of it. All species in this genus have two simple (unsegmented) rays in the small dorsal fin, located in front of the first segmented rays. The body is laterally compressed, short, and tall, about three times as long as it is tall. The two nostrils are very close together, with the posterior one opening near the eye; the body is not very elongated; the odd-numbered fins are sub-equal, with the dorsal fin beginning at the same level or slightly behind the origin of the anal fin; the teeth are bicuspid (Lavoué, 2001). According to Lavoué et *al.* (2004), thirty species and subspecies are known, of which only one species was sampled in our study.

Species Key

Narrow mouth, width ranging from 21 to 25% of total length; fewer than 24 teeth on the lower jaw; presence of a black spot on the flanks, beneath the first rays of the dorsal fin, and at the base of the caudal fin *P. christyi.*

3.2.1.2. Cyprinidae family

Fish in this family have a compressed body, covered with cycloid scales, and a bare head. The mouth is protractile and toothless. The pharyngeal teeth are arranged in 1–3 rows but never exceed 8 teeth per row (Nelson, 2006). They have either one or two pairs of barbels. The lateral line is visible, extending from the posterior part of the operculum to the base of the caudal fin. The fins are composed of jointed rays; one ray of the dorsal fin may be ossified, forming a more or less developed spine. The pelvic fins are inserted in the middle of the abdomen. From this family, only two genera were captured during this study: *Barbus* and *Labeo.*

Key to genera

1-Scales of the mid-lateral row not enlarged or slightly wider than the scales of the adjacent series, 7 to 8 branched rays in the dorsal fin .... *Barbus*

-Specialized mouth with lips lacking a horny covering and not forming a ventral sucker 2

2-Mouth distinctly inferior with well-developed lips, no sucker-shaped chin disc *Labeo*

***Barbus* Cuvier and Cloquet, 1816**

De Weirdt et *al.* (2007) report that the genus Barbus, also known as the "small Barbus group," includes small to medium-sized fish (generally not exceeding 200 mm).

A black spot is located behind the operculum, and another carmine-red spot is situated above the posterior corner of the opercular slit. There are three black spots at the base of the dorsal fin. When preserved, the coloration is dark brown; the band and spots remain indistinct.

***Labeo* Cuvier, 1814**

This genus generally includes fish with a cylindrical body covered with large scales. The head is nearly flattened dorsoventrally. The eyes are nearly dorsolateral. The mouth is sucker-like. Each lip is equipped on the inner surface with small papillae forming rows of transverse folds (Tshibwabwa, 1997). The dorsal fin is spiny, comprising 12 or more rays and beginning in front of the ventral fins. The anal fin is short, consisting of 7 or 8 rays ( ). The intestine is long and highly convoluted. These fish are herbivorous (Tshibwabwa and Teugels, 1995; Daget, 1954; Reid, 1985).

Species Key

-Snout with a transverse groove and bearing a fleshy, upward-pointing appendage *L. weekssi.*

***Labeo weeksii* Boulenger, 1909**

Observed characteristics

Of a whitish, almost silvery coloration, the body is elongated, with a height ranging between 20 and 21% of total length (TL). The head is dorsoventrally flattened and measures 22–25% of TL. The snout is prominent. A broad and deep transverse groove is present. The mouth is wide with wrinkled lips on the inner surface. Maxillary and rostral barbels are present, but are tiny and hidden. The eye is superolateral; its diameter ranges from 19 to 23% of the head length. The length of the caudal peduncle is 6–9% and its height is 17–18% of the standard length.

Description

These fish have a fusiform or elongated body with a height ranging from 19 to 21% of head length. The head measures 27–29% of head length. The snout is sucker-shaped. A large mouth with wrinkled lips on the inside (Tshibwabwa, 1997). Maxillary barbels are present. The eye is superolateral, with a diameter ranging from 20 to 24% of head length. The length of the caudal peduncle is 7–10% and its height 16–17% of head length. The dorsal fin, falciform, consists of 4 simple rays and 10–11 branched rays. The ventral fins extend slightly beyond the genital opening at their tips. The anal fin has a tip that extends beyond the scaly base of the caudal fin. The caudal fin is deeply notched.

Geographic Range

According to Tshibwabwa (1997), this species is widely distributed in the Congo Basin.

Species key

-3rd dorsal ray weakly ossified, not serrated, and flexible .... *Beutonia*

-14 to 16 scales around the caudal peduncle .... *B. cardozoi*

Barbus cardozoi Boulenger, 1912

Observed characteristics

The distance between the tip of the snout and the origin of the dorsal fin ranges from 39 to 50% of total body length (TB). The pectoral fins measure 19–24% of TB and do not reach the ventral fins. The ventral fins measure 16–23% of TB. The anal fin consists of 3 simple rays and 5 branched rays; it does not reach the caudal fin. The caudal fin is notched.

The body width ranges from 49 to 60% of the head length. The eye diameter ranges from 31 to 40% of the head length. The snout length ranges from 22 to 29% of the head length. The mouth is terminal; its width is 17–26% of the head length. The lower jaw is slightly shorter than the upper jaw. The lips are poorly developed. Two pairs of barbels are present: the anterior pair measures 35–41% of the head length and the posterior pair 35–49%.

There are 4–5 gill rakers on the underside of the first gill arch. The caudal peduncle is longer than it is tall. The lateral line consists of 28–33 scales, with 4–6 between the dorsal fin and the lateral line, and 13–16 scales around the caudal peduncle. The dorsal fin, with a concave edge, consists of 2 to 3 simple rays and 6 branched rays. The general body coloration is white or silvery, with the scales on the flanks bearing slightly dark spots. Specimens preserved in formalin have a yellowish coloration.

Description

The body height ranges from 26 to 30% of total body length (TL), with a width ranging from 53 to 63% of head length. The eye diameter ranges from 34 to 42% of head length. The interorbital distance is 25–33% of the head length. The snout length ranges from 25 to 31% of the head length. The mouth is terminal; its width is 18–27% of the head length. The lower jaw is slightly shorter than the upper jaw. The lips are poorly developed. Two pairs of barbels are present: the anterior pair measures 35–41% of the head length and the posterior pair 36–51%.

On the underside of the first gill arch, there are 5 gill spines. The caudal peduncle is 150–182% as long as it is high. The lateral line consists of 29–35 scales, 5–6 between the dorsal fin and the lateral line, 4–5 between the lateral line and the midpoint of the belly, 3 between the lateral line and the origin of the ventral fin, 10–12 predorsal scales, and 14–16 around the caudal peduncle. The dorsal fin, with a concave margin, consists of 3 unbranched rays and 7 branched rays. The third unbranched ray is ossified and serrated; the longest ray measures 18–25% LS. The origin of the dorsal fin lies on the same vertical line as that of the ventral fins.

Between the tip of the snout and the origin of the dorsal fin, we measured 42–49% LS. The pectoral fins measure 20–23% LS and do not reach the ventral fin. The ventral fins measure 17–22% LS. The anal fin consists of 3 simple rays and 5 branched rays; it does not reach the caudal fin. The caudal fin is notched (Skelton, 2001).

Geographic range

Lévêque et *al.* (1984) state that this species is known from many countries in Southeast Africa.

Barbus eutonia Boulenger, 1904

Observed characteristics

With a body height ranging from 28 to 34% of total body length (TL), the fish has a head measuring 25–30% of TL and a width of 52–60% of head length. The snout width ranges from 21 to 29% of head length. The mouth has a lower jaw that is shorter than the upper jaw.

There are two pairs of barbels: the anterior pair measures 62–89% of the head length, and the posterior pair 33–45%. The lateral line consists of 20–22 scales. The dorsal fin consists of 2 to 3 (rarely 4) unbranched rays and 5 to 7 branched rays. The anal fin consists of 2 to 3 unbranched rays and 5 to 6 branched rays. The pectoral fins do not reach the ventral fins. The body of living specimens has a yellowish coloration. When preserved in formalin, the body takes on a brownish coloration.

Description

Body height ranges from 27 to 34% of total body length (TBL). The head measures 27–34% of TBL. Its width is 54–68% of head length. Eye diameter ranges from 26 to 36% of head length. Interorbital distance measures 28–36% of head length. The snout is subconical; its length ranges from 25 to 34% of head length. The mouth is terminal, and the lower jaw is shorter than the upper jaw (Skelton, 2001).

The lips are poorly developed; the lower lip measures 35–45% of the head length. Two pairs of barbels are present: the anterior pair measures 64–92% of the head length and the posterior pair 36–59%. The caudal peduncle is 173–224% as long as it is high. The lateral line consists of 24–27 scales; 4.5 between the dorsal fin and the lateral line; 4.5 between the lateral line and the midpoint of the belly; 2–3 between the lateral line and the ventral fin; and 12 around the caudal peduncle. The dorsal fin, with a concave edge, consists of 3 unbranched rays and 6 branched rays. The third unbranched ray is ossified and serrated; the longest ray measures 41–54% of the head length. The anal fin consists of 3 unbranched rays and 6 branched rays; it does not reach the caudal fin. The pectoral fins do not reach the ventral fins, and the caudal fin is notched.

Geographic range

According to Lévêque et *al.* (1984), the species is known from the Congo Basin.

3.2.1.3. Family Citharinidae

According to Gosse et *al.* (2003), members of the family Citharinidae have a very high, laterally compressed body covered with cycloid scales (*Citharinops* and *Citharinus*) and ctenoid scales (*Citharidium).* The terminal mouth has a row of small monocuspid teeth along the edge of the lips. In contrast, the maxilla, which is very reduced, lacks teeth. All species have a fairly large adipose dorsal fin located behind the rayed dorsal fin (16 to 24 branched soft rays). The anal fin has 19 to 24 branched soft rays. The lateral line is complete, straight, and median (47 to 92 scales). All species, which are highly specialized microphages, possess numerous fine and very closely spaced gill spines. Only one species was captured during this study.

***Citharinus* Cuvier, 1817.**

Citharinus are distinguished by the combination of the following characteristics: body covered with small cycloid scales, more than 60 scales in the longitudinal line, more than 12 scales between the lateral line and the base of the ventral fins (Gosse et *al.,* 2003).

61 to 64 scales in the lateral line; 16.5 to 15.5 scales between the lateral line and the origin of the dorsal fin; very small adipose fin, its base measuring 2.5 to 2.2 times the distance separating it from the dorsal fin *C. congicus*

***Citharinus congicus* Boulenger, 1897**

Characteristics observed

Presence of a very small adipose fin whose base is 3.1 times the distance separating it from the dorsal fin; a high body (body height 45.9% of LS). The snout is short, broad, and slightly pointed. The dorsal fin, which has 18 rays, is elevated at the front and falcate. The anal fin has 28 to 30 rays. There are 65 lateral line scales, 17 scales between the lateral line and the origin of the dorsal fin, and 21 scales around the caudal peduncle. There are 28 gill rakers on the lower part of the first gill arch. Living specimens have a silvery or whitish coloration.

Description

The body height of the species is 1.8 to 2 times the total length. The dorsal fin has 18 to 21 rays, including 4 to 5 unbranched rays; the anal fin has 25 to 30 rays, including 3 unbranched rays. The number of lateral line scales ranges from 60 to 66.15, with 16 scales between the lateral line and the origin of the dorsal fin. There are 30 to 40 gill rakers on the lower part of the first gill arch. The adipose fin extends 2.1 to 2.7 times the distance between it and the dorsal fin (Boulenger, 1909; Poll, 1976).

Geographic range

The species is widely distributed throughout the Congo Basin (Poll, 1976) and in the Ruaha River in Tanzania (Daget, 1984b).

3.2.1.4. Family Distichodontidae

This family is endemic to freshwater systems throughout sub-Saharan Africa and along the Nile. Members of the family exhibit a remarkable range of morphologies, particularly in their jaws, which reflect the diversity of their feeding habits. Members of this family are easily distinguished from other African characins by the combination of ctenoid scales with ctenii formed by a series of independent ossifications along the posterior edge of the scale and by having at least a few bicuspid teeth on each jaw (Vari, 2007). In addition to these characteristics, Gosse et *al.* (2003) report that members of this family have a long body (longer in *Distichodus),* a generally present adipose fin, a median lateral line, and well-developed teeth. Daget and Gosse (1984) recognized 17 genera comprising 90 species. In the study area, only one genus and one species are present.

Genus key

Lateral line scales perforated along the entire length of the body, lower jaw with a movable joint between the dentary and the anguloarticular, maxilla lacking teeth..................*Distichodus*

Distichodus Muller and Troschel, 1845

Distichodus is distinguished by the combination of the following characters: an upper jaw that is slightly mobile relative to the skull, a toothless maxilla not closely attached to the posterior part of the premaxilla, ctenoid scales, a movable joint between the dentary and the anguloarticular, and a completely perforated lateral line. 24 species are currently valid (Vari, 2007). Currently, taking into account the recent description *of D. teugelsi* from the Middle Congo by Mamonekene and Vreven (2008), this genus now comprises 26 species .

Species Key

20 to 26 rays in the dorsal fin; absence of rounded blackish spots on the flanks, 17 to 19 small dark vertical bands across the body, which barely reach the belly......*D. fasciolatus*

***Distichodus fasciolatus* Boulenger, 1899**

Observed characteristics

Distichodus fasciolatus is notable for its numerous small dark bands across the body (17 to 20), which rarely extend to the belly. The dorsal fin has 25 to 26 rays, and the anal fin has 13 to 15 rays. The forked caudal fin, almost entirely covered with small ctenoid scales, has a pointed upper lobe and a rounded or obtusely pointed lower lobe. The scales, strongly ctenoid, number 70 to 76 along the lateral line, with 25 scales around the caudal peduncle. There are 13 to 15 gill rakers on the lower part of the first gill arch. Live specimens are generally yellowish-gray, darker dorsally. The odd-numbered fins, which tend to be yellowish, are slightly blackened; the dorsal fin has numerous black spots, and the caudal fin is edged with black posteriorly. The even-numbered fins are pale yellow.

Description

The body height of the species ranges from 2.7 to 3 times the standard length (SL). The head length is 4.5 to 5 times the standard length (SL). The dorsal fin has 23 to 28 rays, and the anal fin has 13 to 16 rays. The anal fin has 13 to 16 rays. The number of scales ranges from 68 to 78 along the lateral line, with 24 scales around the caudal peduncle. There are 13 to 15 gill spines on the lower part of the first gill arch (Boulenger, 1909; Poll, 1967a).

Geographic range

Distichodus fasciolatus is known from the Congo Basin (Daget et *al.,* 1984).

3.2.1.5. Family Alestidae

This large African family was formerly classified as "Alestinae" within the family Characidae (Greenwood *et al.,* 1966). According to Paugy and Schaefer (2007), the family Alestidae comprises approximately 110 small- to medium-sized species, widely distributed throughout most of the basins of intertropical and sub-Saharan Africa. Alestidae exhibit varied forms and a fairly diverse ecology. The main characteristic of this family is strong dentition. However, the number and shape of the teeth vary and serve as good criteria for identifying the main families (Paugy, 2003c).

Key characteristic of the genus

Presence of a single row of teeth on the upper jaw ....*Hydrocynus*

Hydrocynus Cuvier, 1817

Hydrocynus species look virtually identical, and only a close examination can distinguish them. They have a slender shape, likely related to their feeding habits, as they chase their prey—hence the name "tigerfish." Their overall coloration is silvery and shiny. The scales are marked with a dark spot, forming parallel bands that are most visible above the lateral line. In some species, these lines are more or less dark. The dorsal fin is positioned at the same level as the ventral fins or slightly forward. The mouth contains a single row of highly developed, sharp teeth on each jaw (10–13/12) (Paugy *et al.,* 2007). A single species has been collected from the Lubilanji and Luilu rivers.

Species Key

The tip of the dorsal fin is black-striped; the notch in the caudal fin is bordered by black; the adipose fin is black; the black-striped dorsal fin is inserted at approximately the same level as the ventral fins *Hydrocynus vittatus.*

***Hydrocynus vittatus* Castelnau, 1867**

Characteristics observed

This species has 38 to 52 scales along the lateral line and 6 to 8 scales between the lateral line and the origin of the dorsal fin. The dorsal fin is inserted very slightly forward. The eye occupies at least 68% of the interorbital space. The gill spines are few in number, 5 to 8 on the lower part of the first gill arch and 10 to 12 on the upper part. The body of *H. vittatus* is shiny silver, bearing well-defined dark longitudinal lines along each scale row. The pectoral, pelvic, and anal fins are yellowish-gray.

Description

The species has 43 to 53 scales along the lateral line and 7.5 scales between the lateral line and the origin of the dorsal fin. The dorsal fin is inserted very slightly in front of or at the same level as the ventral fins. The eye occupies at least 70.0% of the interorbital space. The gill rakers are few in number, 5 to 9 on the lower part of the first gill arch and 9 to 12 on the upper part (Paugy et *al.,* 2007).

Geographic range

The species is reported from: Cross, Sanaga, Chad, Niger/Bénoué, Ouémé, Senegal, Nile, Omo, the entire Congo Basin, Lufira, Lualaba, Lake Bangwéolo, Luapula, Lake

Moéro, Lake Tanganyika, Lake Upemba, Zambezi, Limpopo, Rovuma, Lake Rukwa, Shire, Rufiji, and Ruaha (Paugy, 1984, 2003c; Paugy *et al.,* 2007).

3.2.1.6. Family Bagridae (Claroteidae)

These are catfish whose nostrils are located on the snout; presence of a nasal barb on the anterior nostrils. The caudal fin is fairly deeply forked; fewer than 7 branched rays in the dorsal fin (Greerinckx *et al.,* 2007). Only one genus was recorded in our study.

***Chrysicthys* Bleeker, 1858**

The genus *Chrysicthys* is characterized by the presence of four pairs of barbels; a dorsal fin with 5 to 7 soft rays, preceded by a very short spine and a strongly developed spine that is slightly serrated along its posterior margin; a medium-sized or small adipose fin (the base being smaller than the width of the head) that is never ossified; a pair of pectoral fins with 8 to 11 soft rays, preceded by a strong spine, well-serrated along the posterior margin; a pair of ventral fins, located approximately in the middle of the body, with 5 rays. The anal fin, of medium size, has 3 to 6 unbranched rays and 6 to 12 branched rays. The caudal fin is well forked; the eyes, with free margins, are lateral and large; and the body is moderately elongated, 4 to 6 times longer than high (Risch, 2003).

Key to the genus

Head and body not flattened; head without lateral bulges or notable skin appendages; presence of 4 pairs of barbels *Chrysicthys*

Species key

1-Body uniformly colored, 8 to 12 soft rays in the pectoral fin... .*C. duttoni*

-Body bearing small dark brown to black spots or typical mottling, 7 to 8 soft rays in the pectoral fin 2

2-Presence of small dark brown to black spots or patches or typical mottling on the body, 7 to 8 soft rays in the pectoral fin, body bearing small dark brown to black spots or patches, mouth inferior *C. auratus*

-Presence of typical mottling on the body, terminal mouth, lower jaw often prominent, skull well-developed and visible through the skin with 8 soft rays in the pectoral fin...................................*C. ornatus*

Chrysicthys duttoni Steindachner, 1905

Characteristics observed

Presence of four pairs of barbels; a dorsal fin with 5 soft rays, preceded by a very short spine and a strongly developed spine; a medium-sized, non-ossified adipose fin; presence of a pair of pectoral fins with 8 to 9 soft rays. The anal fin, of medium size, has 4 to 6 unbranched rays and 7 to 11 branched rays. The caudal fin is well-forked; the eyes, with free rims, are lateral and large; and the body is moderately elongated, 2 to 6 times longer than high. Body length varies between 15 and 24 cm. Captured specimens are brownish on the back and head, with the rest of the body whitish. The fins are grayish.

Description

Chrysicthys duttoni is characterized by the presence of four pairs of barbels; a dorsal fin with 5 soft rays, preceded by a very short spine and by a strongly developed, weakly denticulated spine ( ) at its posterior margin; a medium-sized or small adipose fin (the base being smaller than the width of the head) that is never ossified; a pair of pectoral fins with 7 to 12 soft rays, preceded by a strong spine, well-serrated along the posterior margin; a pair of ventral fins, located approximately in the middle of the body, with 6 rays. The anal fin, of medium size, has 2 to 8 unbranched rays and 5 to 13 branched rays. The caudal fin is well-forked; the eyes, with free margins, are lateral and large; and the body is moderately elongated, 3 to 7 times longer than high (Risch, 2003; Boulenger, 1916).

Geographic range

According to Risch (1986), *Chrysicthys duttoni* is known throughout tropical Africa, in the Congo Basin, with the exception of most of the upper reaches of the Lualaba and Luapula rivers.

***Chrysicthys auratus* Saint-Hilaire, 1809**

Characteristics observed

Moderately elongated body; body height ranges from 16.9% to 24.7% of total length (TL). The head is slightly flattened, with a length of 31.6% to 34.8% of total length (TL). The eyes are wide, ranging from 17.8% to 32.5% of total length (TL). Specimens have a silvery to brownish coloration.

Description

The body of this species is moderately elongated, with a body height of 18.1 to 26.5% of total body length (TSL). The head is slightly flattened, with a length of 30.3 to 36.1% of TSL. The large eyes have a diameter ranging from 19.5 to 34.6% of total head length (THL) (Risch, 1986).

Geographic range

According to Risch (1986), *Chrysicthys auratus* is known from the entire Congo Basin.

***Chrysicthys ornatus* Boulenger, 1909**

Observed characteristics

This species is distinguished by the coloration of its body, which bears typical marbling, a terminal mouth, an often prominent lower jaw, and a well-developed skull visible through the skin. The dorsal and pectoral fins have 4 to 8 soft rays, and rarely 10. There are 8 to 12 gill spines on the lower part of the first gill arch. Captured *Chrysicthys ornatus* specimens have a brownish coloration. The body, head, and even the fins are mottled.

Description

The body of the species is compressed, with a body height of 15.2 to 22.9% of total length (TL). The long, compressed head measures 31.9 to 37.5% of TL, with eyes whose diameter ranges from 15.7 to 29.0% of head length (Risch, 1986).

Geographic range

According to Risch (1986), *Chrysicthys ornatus* is widely distributed in the Congo Basin.

3.2.1.7. Family Mochokidae

According to Paugy and Roberts (2003), all fish belonging to this family have a scaleless body. They possess three pairs of barbels—one maxillary pair and two mandibular pairs—except in certain rheophilic forms, whose lips are modified into adhesive discs. There are no nasal barbels. The dorsal fin has a spiny anterior ray; the adipose fin is large and sometimes rayed (genus *Mochokus).* The first ray of the pectoral fins is spiny and denticulate. There is a strong cephalonucal shield. According *to* Fermon et *al.* (2007), the family Mochokidae is the most species-rich among catfish and currently comprises 10 genera with over 170 species.

Generic Key

Absence of an adhesive or oral labial disc on the lips, eye with a free margin, mandibular teeth clustered toward the symphysis of the jaw, and the pectoral spine with poorly developed or absent denticles*Synodontis*

***Synodontis* Cuvier, 1817**

Synodontis possess a dorsal fin with 6 or 7 branched rays preceded by a spine that is more or less denticulated on its anterior surface and a fairly large adipose fin. The pectoral fins are equipped with a strong, denticulated bony spine whose teeth are more developed on the inner side than on the outer side. There is a pair of maxillary barbels, which may or may not be membranous, varying in length depending on the species. The two pairs of mandibular barbels are more or less short and always branched. The premaxillary teeth are short and conical; the mandibular teeth, which are mobile, slender, and curved, are set on the lower lip. The body coloration, which is quite varied, is a frequently used identification criterion (Paugy et *al.,* 2003).

Species Key

Distance from snout to base of dorsal fin 41.2–45.8% LS, postorbital distance 50.8–69.3% of snout length, 17–23 mandibular teeth, lower lip more or less dark, mottled with light, ventral regions black *S. nigromaculatus*

Synodontis nigromaculatus Boulenger, 1905

Characteristics observed

This species is distinguished by the distance between the snout and the origin of the dorsal fin, which is 42.3 to 45.4% of total body length. The mouth is small, with non-enlarged lips and few papillae, bearing 16 to 24 short, movable mandibular teeth arranged in a transverse band. The dorsal fin has one spine and 7 to 9 branched rays, and the anal fin has 2 to 4 simple rays and 6 to 9 branched rays. This species exhibits variable coloration;

the dorsal regions and flanks are brownish.

Description

In S. nigromaculatus, the body depth and head length are 26.9–36.7% FL and 29.1–34% FL, respectively. The maxillary barbels, with a distinct basal membrane, measure 40.6–85.8% FL. The distance of the anal fin from the base of the adipose fin ranges from 20.7 to 25.3% SL. The length of the dorsal spine and that of the pectoral spine range from 21.6 to 33.8% SL and 28.3 to 34.3% SL, respectively The subrounded snout measures 45.4 to 55% of total length (TL). The humeral process, whose length varies between 60.7 and 75.1% of TL, is triangular, pointed, granular, and striated, with a slight crest on top. The eyes are superolateral and measure 27.7 to 37.3% of TL. The slightly concave interorbital space ranges from 29.9 to 36.2% of total length. The external and internal mandibular barbels range from 66.7 to 133.5% and 34.1 to 61% of total length, respectively (Poll, 1971).

Geographic range

Synodontis nigromaculatus has been reported from the central Congo Basin (Poll,

1971).

3.2.1.8. Family Auchenoglanididae

This family of catfish was classified as a subfamily of the Claroteidae by Geerinckx et *al.* (2007). According to these authors, fish in this family have anterior nares on the upper lip, and posterior nares without a nasal barb, or with a rudimentary barb of varying length (up to 3 mm). The caudal fin lacks a slight notch; the dorsal fin has 7 or more branched rays. Six genera and approximately 28 species are present in this family (Geerinckx et al. 2004).

Genus Key

Posterior cranial bones clearly visible through the skin, large supraoccipital process, and well-developed first nuchal plate in contact (or even overlapping). The premaxillary tooth band is small (total width 5–8% of total length). The edge of the caudal fin is slightly concave or straight *Auchenoglanis*

***Auchenoglanis* Gunther, 1865**

Auchenoglanis can be distinguished from related genera by its pointed snout, steep skull, and high body. However, it is the parieto-supraorbital bone in contact with the strongly developed first nuchal plate that characterizes this genus. The premaxillary tooth band is small (total width 4–7% TL) (Geerinckx *et al.,* 2004).

***Auchenoglanis occidentalis* (Valenciennes, 1840)**

Observed characteristics

This species has a high, stiff head with a long, pointed snout; the length in front of the eyes measures more than half the head length in most cases. The maxillary barbels are shorter than the outer mandibular barbels and do not extend beyond the eye. The width of the premaxillary band measures less than 6 to 7% of total length (TL). The adipose fin is high and rounded. The dorsal fin has 5 to 7 soft, branched dorsal rays.

Specimens exhibited dark spots on the flanks and fins.

Description

Auchenoglanis occidentalis has a high, stiff head with a long, pointed snout; the length in front of the eyes measures more than half the head length in most cases. The maxillary barbels are shorter than the external mandibular barbels and do not extend beyond the eye. The width of the premaxillary band ( ) measures less than 10% of total length (TL). The adipose fin is high and rounded. The dorsal fin has 7 soft, branched dorsal rays (Geerinckx *et al.,* 2007).

Geographic range

According to Geerinckx *et al.* (2007), *Auchenoglanis occidentalis* has a wide geographic distribution: the species is found in the rivers of West Africa, in the Chad Basin, in the Cross River, in northwestern Cameroon, as well as in the Nile, throughout the Congo, and in Lake Tanganyika.

3.2.1.9. Clariidae family

Fish in this family have an eel-like body. They are commonly known as catfish. They possess four pairs of barbels. They have a suprabranchial organ, formed by evaginations of the second and fourth gill arches, which allows the fish to breathe air. This organ is a synapomorphy for the family.

The dorsal and anal fins are very long. The dorsal fin lacks spines. The caudal fin is rounded. The family comprises 14 genera and 90 species (Teugels *et al.*, 2003).

2.3.3.1. *Clarias angolensis* Steindachner, 1866

Specimens are characteristic with a uniformly dark brown coloration. The pectoral spine is strongly dentate on both sides.

The neuromasters on the flanks display an irregular pattern. The maxillary barbels barely exceed the length of the head. Head length is 21.8 to 25.9% of total length (TL), and cleithrum width ranges from 12.7 to 18.2% of TL. There are 75 to 82 rays in the dorsal fin and 53 to 60 rays in the anal fin. The gill rakers on the first gill arch range from 14 to 24. Captured fish have dark brown and light brown coloration on the belly.

The species is characterized by a strongly dentate pectoral spine on both sides. The neuromasters on the flanks exhibit an irregular pattern. Barbell length (length of nasal barbels 72.1 to 130.2% TL; length of maxillary barbels 105.1 to 195.3% TL; length of internal mandibular barbels 72.4 to 108.9% TL; length of external mandibular barbels 95 to 134.5% TL) and by the relatively narrow dental bands (width of the premaxillary band 24.3 to 30.7% TL; width of the vomerine band 20.9 to 27.5% TL). The length of the claeithrum ranges from 13.0 to 18.6% TL. According to Teugels (1986), head length ranges from 22.3 to 27.1% SL, and head width from 14.1 to 19.4% TL. The dorsal fin has 67 to 84 rays and the anal fin 55 to 59 rays. The number of gill rakers on the first gill arch ranges from 10 to 23 (Teugels *op. cit.).*

This species is found in the lower reaches of the Ogôoué River in Gabon (Mbega, 2004), in the Kouilou River in the Republic of the Congo, and in the Lower and Middle Congo (Teugels, 1986).

I.3.3.2. *Clarias gariepinus* Burchell, 1822

The coloration of the entire body is mottled with black spots; the belly is whitish- , and the lower part of the head shows two blackish lateral bands. The head is very flattened. Head length is 29.1–32.9% of total body length (TBL), and head width is 18.1–23.2% of TBL. The pectoral spine is denticulate on the anterior portion. There are 63 to 78 soft rays in the dorsal fin and 45 to 61 in the anal fin. The gill rakers number 31 to 56 across the entire first gill arch.

Measuring 25–36% of total body length, the head is more or less rectangular in shape. The small eyes are positioned superolaterally and measure 6–13% of head length. The snout is long and broad (Teugels *et al.*, 2007). The number of gill rakers on the first gill arch varies between 24 and 110.

Its pectoral spine is hard, serrated on the anterior part and smooth on the posterior part. The distance between the tip of the dorsal fin and the caudal fin is not great (Teugels, *op. cit.*).

This species has a wide distribution; it is found in most rivers of the Congo Basin. In the central basin, it has been found on several occasions.

3.2.1.10. Family Schilbeidae

The Schilbeidae (a family of catfish found in Africa and Asia); they have a fusiform body with a broad, flattened head and lateral eyes. Fish in this family have 3–4 pairs of barbels at the mouth (De Vos, 1995). The dorsal fin is short, very anterior, and preceded by a spine; in other species of this family, it is absent. The adipose fin is small, well separated from the rayed dorsal fin, and sometimes absent. The anal fin is very long, located beneath the caudal peduncle and extending to the caudal fin. A dark, rounded humeral "spot" behind the head where the swim bladder contacts the skin. Currently, there are 5 genera and 33 species.

Genus Key

Two pairs of internal mandibular barbels present, adipose fin absent or rudimentary, dorsal fin with six branched rays..*Schilbe*

***Schilbe* Oken, 1817**

A fin with 5–6 soft rays; the dorsal and pectoral fins consisting of a spine; the adipose fin, which is sometimes absent, are characteristic of species in this family. The ventral fin has a single spine and 5 other soft rays. They have 4 pairs of barbels (De Vos 1995).

Species key

1-Adipose fin absent...*S. mystus*

-Adipose fin present 2

2-Absence of dark lateral bands; maxillary barbels not extending beyond the posterior edge of the operculum *S. grenfelli*

***Schilbe mystus* Boulenger, 1900**

Observed characteristics

The body height measures 18–33% of total body length (TL), and the head length 15–25%. The snout is somewhat long. The mouth width ranges from 55 to 64% of the head length. Nasal barbels are present; the maxillary barbels extend at least to the anterior margin of the operculum and at most to its posterior margin; the gill spines are short, widely spaced, and number 6–12 on the first gill arch. The coloration is dark gray, with a silvery belly.

Description

Body height measures 19–35% of total length (TL), and head length 19–26%. The head is 133–152% as long as it is wide. The snout is slightly longer than the eye and measures 24–39% of head length. Eye diameter ranges from 19 to 29% of head length. The width of the mouth is between 56 and 66% of the head length. The nasal barbels reach at least the anterior edge of the eye and at most its posterior edge. The maxillary barbels reach at least the anterior edge of the operculum and at most its posterior edge; the external mandibular barbels measure 23–42%, and the internal mandibular barbels are short (13–25% of head length) and never reach the posterior edge of the eye. The gill spines are short, widely spaced, and number 9–14 on the first gill arch (De Vos, 1995).

Geographic range

According to De Vos (1995), the species is widespread throughout the Congo Basin.

Schilbe grenfelli Boulenger, 1900

Characteristics observed

Schilbe grenfelli is distinguished by the presence of an adipose fin. The anterior nares are closer together than the posterior nares. The head is rounded with a prominent snout. The mouth is subterminal. The pectoral spines are weakly dentate on the posterior portion. There are 45 to 53 branched rays in the anal fin and 6 to 9 gill spines below the first gill arch. The nasal barbels do not extend beyond the posterior edge of the eye. Specimens range in color from dark to silvery.

Description

The body height of the species ranges from 19.0 to 32.9% of total body length (TL), and the head length from 17.9 to 25.3% of TL. The anal fin has 44 to 59 rays. There are 7 to 11 gill rakers on the lower part and 2 to 4 on the upper part of the first gill arch. The species has 46 to 49 unfused vertebrae (De Vos, 1995).

Geographic range

Schilbe grenfelli is also widespread throughout the Congo Basin.

3.2.1.11. Family Malapteruridae

The Malapteruridae are a group of electric catfish easily distinguished by their thick , and fleshy bodies, with small, laterally positioned eyes. They lack scales and spiny rays on their fins. They lack rayed dorsal fins. They have a relatively long adipose fin located near the caudal fin and an organ in the abdominal cavity that produces powerful electric discharges. The Malapteruridae family is endemic to Africa, with uncertain phylogenetic relationships (Norris, 2007). According to Norris (2002), 19 species divided into 2 genera are known in this family.

***Malapterurus* Lacépède, 1803**

Malapterurus species reach at least 15% of their standard length (SL) at maturity. They possess a two-chambered swim bladder. Observations regarding the lifestyle of these fish are largely anecdotal, but most species appear to be piscivorous, using electric discharges to stun their prey (and also to ward off predators). Malapterurus is found in a variety of aquatic habitats, sometimes living (and likely breeding) in shelters along riverbanks (Norris, 2007).

***Malapterurus electricus* Poll and Gosse, 1669**

Observed characteristics

This species has a relatively fusiform body, a flattened head, and a narrow snout and dental plates. The pectoral fin rays number 8 to 9, the pelvic fin rays 6, and the anal fin rays 11 to 12. There are 19 caudal rays. The gill spines on the upper part of the gill arch are fewer in number (2 to 3), while those on the lower part number 11 to 12. The captured *Malapterurus electricus* is brownish, darker dorsally, much lighter ventrally, and dotted with black spots of varying sizes, which are rare or absent on the belly. All fins are also more or less covered with smaller spots that are more or less densely clustered.

Description

In this species, the total number of gill rakers on the first gill arch ranges from 7 to 29, the number of vertebrae ranges from 27 to 28, the number of caudal vertebrae from 23 to 25, and the total number of vertebrae from 40 to 42. The rays of the pectoral fins range from 8 to 10, and those of the anal fin from 10 to 13 (Norris, 2002).

Geographic Range

This species is known from the Central Congo Basin (Norris, 2002), the Nile, and tropical Africa, including the Great Lakes with the exception of Lake Victoria, and the rivers of East Africa north of the Zambezi.

3.2.1.12. Family Cichlidae

According *to* Stiassny *et al.* (2007), these fish have a compressed body that is completely covered with ctenoid and cycloid scales. The lower pharyngeal bones, fused together, form a serrated triangle. They have a pair of nostrils. The lateral line is interrupted. The dorsal fin consists of a spiny portion and a soft portion. These same authors noted that new species of this family are discovered every year.

Key to the genera

I-Cycloid flank scales 2

2-Lower lateral line extends over the anterior part of the flank well beyond the end of the upper lateral line; trifid at the rear of the caudal fin. The first spine of the anal fin is very short; the second and third spines are long and of equal size .... *Tylochromis*

-Lower lateral line does not extend beyond the tip of the upper lateral line and is not trifid at the rear 3

3-The first three spines of the anal fin increase gradually in length; at least 15 long gill spines on the lower part of the first gill arch

gill *Pterochromis*

-Approximately 17 spines on the dorsal fin... 4

4-Lower pharyngeal bone as long as it is wide, with large holes at the lower base; 8 to 13 gill spines on the lower part of the first gill arch; robust, immobile teeth; presence of a black spot on the soft part of the dorsal fin (in juveniles and many adults) ....*Tilapia*

5-The lower part of the first gill arch has more than 19 to 21 gill rakers 6.

6-Caudal fin striped with black bands alternating with light bands; 14 to 26 gill rakers on the lower part of the first gill arch, fine and mobile teeth *Oreochromis*

***Oreochromis macrochir* (Boulenger, 1912)**

Description**:** Elevated body with a rounded snout and a high caudal peduncle. Head length 31.2–38.3% of total body length. Jaws with 4–7 rows of teeth with fairly slender bodies and moderately widened crowns; outer row of bicuspid teeth; in large females, presence of a few tricuspid teeth; in large males, some teeth worn down to become unicuspid. Lower pharyngeal bone with very fine teeth; posterior teeth with a well-marked shoulder.

Toothed zone of the lower pharyngeal bone with broad, rounded lobes; ventral keel elongated, longer than the toothed portion. 20–26, usually 22–24, gill spines on the lower part of the first gill arch in specimens 90 mm TL or larger. Dorsal fin with 15–17 spines and 11–14 soft rays; anal fin with 3 spines and 9–12 soft rays. The long pectoral fins extend vertically from the base of the anal fin. Cycloid scales, 31–32 along the lateral line,

Cheeks with 2–3 rows of scales. Maximum length: 400 mm TL. Coloration: silvery juveniles with 8–10 narrow, optional dark vertical bars on the body and yellowish fins. If present on preserved specimens, a tilapia-like spot on the dorsal fin, but this is at most a darker portion of any other pattern (bars or reticulation). Caudal fin with a few dark spots in the center or unmarked.

Adult females and non-breeding males are pale green, silvery, or buff with a white or yellowish belly, sometimes with dark centers on the scales giving the appearance of longitudinal stripes. Head, darker green; snout , iridescent blue-green.

No mid-lateral spots. The color of the lower jaw and gill cover varies; it may be unpigmented or show an irregular and variable distribution of melanin in transverse stripes beneath the chin.

The operculum is slightly marbled with a dark green or iridescent spot; on the front of the head, a few dark spots, generally associated with the pores of the lateral line system. Caudal fin with a few dark spots, or darker between the rays. Dorsal fin with a yellow or partially pale red edge and green spots on the soft membranes.

Breeding males, generally larger than females, have a very dark green, almost black, iridescent body and vertical fins with a bright red edge on the spiny and soft dorsal fins and the caudal fin. They possess a prominent white, tuft-shaped genital papilla with numerous orange or yellow spots (drops).

Natural distribution: known from the Upper Zambezi, the Okavango and Ngami regions, the Cunene Basin, the Kafue River, the Chambezi River, and the Bangweulu region.

***Oreochromis mossambicus* (Peters, 1852)**

Description: Moderately elevated body, body depth 36.0–49.5% of total length (TL). Head length 32.3–37.0% of TL in females, 34.0–39.0% of TL in males. Jaws of older males are wider, and the teeth project forward.

Jaws with 3–5 rows of slender teeth; the outer row has bicuspid teeth in juveniles, while all inner rows have tricuspid teeth; as the fish grows, some of the outer teeth are replaced by unicuspid teeth, whereas in very large specimens all outer teeth are simple, and the inner teeth are sometimes simple as well.

Width of the lower pharyngeal bone generally less than its length; ventral keel as long as or longer than the toothed portion. Pharyngeal teeth very slender and pointed; toothed portion with concave margins and fairly narrow lobes.

14–20 branchiospines on the lower part of the first gill arch. Dorsal fin with 15–17 spines and 10–13 soft rays; anal fin with 3 spines and 9–12 soft rays. Cycloid scales; 30–32 along the lateral line, cheeks with 2–3 rows of scales. Maximum size: reaches 400 mm SL.

Coloration: Juveniles are silvery, with 6–7 vertical bars and 3 spots on the flanks. A tilapia-like spot with a clear ring is present up to a size of approximately 80 mm LS. Adult females and non-breeding males lack stripes, but may have the intersection points of the optional stripes represented by 3 to 4 upper spots and 2 to 5 mid-lateral spots. Background color of adults: silvery olive to deep blue-gray.

Dorsal and caudal fins with red edges. Breeding males turn deep blue-black to completely black, with a white lower head and throat, scarlet edges on the dorsal and caudal fins, and red pectoral rays.

Natural distribution: Lower Zambezi, Lower Shire, and coastal plains from the Zambezi Delta to Algoa Bay . Introduction to Lower Guinea: species introduced for aquaculture purposes in 1953 from South Africa to the Republic of the Congo.

According to the FAO (2005), the species reproduces and has established itself in the wild. Its presence in Lower Guinea is not confirmed by museum specimens.

***Oreochromis niloticus* (Linnaeus, 1758)**

Description: Compressed body, height 40–47% of total length (TL). Head length 31.5–40.5% of TL. Jaws with 3–4 rows of teeth, 4–6 rows in specimens over 200 mm TL; the outermost row with bicuspid, stout teeth. Width of the lower pharyngeal bone equal to or greater than its length; ventral keel longer than the toothed region. Lower pharyngeal bone with robust teeth arranged in a more or less triangular zone. 18–28 branchiospines on the lower part of the first gill arch. Dorsal fin with 15–18 spines and 11–14 soft rays; anal fin with 3 spines and 9–11 soft rays. Pectoral fins 33.0–43.5% of total body length (TB). Length of caudal peduncle 10–12% of TB. Cycloid scales; 31–34 along the lateral line; cheeks with 2–3 rows of scales. Maximum size**:** reaches 500 mm total length (TL).

Coloration: This species is easily recognized by the regular vertical stripes on the caudal fin. The tilapia spot is absent in adults but very distinct in juveniles, which also have clearer transverse stripes and a very prominent black spot on the upper part of the caudal peduncle. Adults have a relatively dark grayish background color.

Back olive-green, flanks paler, with 6–9 indistinct transverse bars; belly whitish. Upper lip pale green or white, lower lip white. Dorsal and anal fins grayish, sometimes with a very fine red edge; soft part of the fin with vertical lines or aligned light spots resembling a zebra pattern. Pelvic fins gray, pectoral fins transparent. Throat, belly, and odd-numbered fins black in mature males. Breeding males with a red sheen on the lower parts of the head and body, as well as on the dorsal and caudal fins.

Natural distribution: the Nile basin from the delta to Lake Albert (excluded), the Lake Chad basin, and the Niger, Benue, Volta, Gambia, and Senegal rivers in West Africa. Introduction to Lower Guinea: according to Thys van den Audenaerde (1966), the species was introduced, mainly after 1955, from Fort-Lamy, Chad (12°07' N-15°03' E), to Yaoundé, Cameroon (3°52' N-11°31' E), but Moreau *et al.* (1988) and Welcomme (1985) cite the year of introduction, with unknown origin.

In both cases, the purpose was aquaculture. The species was also introduced in 1975 from Bangui, Central African Republic, to Cameroon. It reproduces and has established itself in the wild, specifically in the Noun marshes and the Djerem and Sanaga rivers, where it has already replaced native species. In Cameroon, it is currently the main species in fish farming and is used throughout the country.

This species was also introduced for aquaculture purposes in 1953 from Sudan to the Republic of the Congo. The species reproduces and has become established in the wild. Furthermore, it is cited by the FAO (1999, 2005) for fish farming purposes in Gabon, having been introduced at an unknown date. It is not known whether the species has become established there or not.

The FAO (1999) reports that aquaculture production in Gabon was 2 tons in 1988, rising to 5 tons between 1992 and 1994, but declining to 3 tons in 1997. Its presence in Lower Guinea is confirmed by museum specimens from Cameroon and Gabon.

No collections are available from the Republic of the Congo. However, Teugels *et al.* (1991) in Daget *et al.* (1991) observed the species among catches by local fishermen at Lake Nanga (4°13' S–11°49' E), Kouilou Basin, and attribute its presence to an introduction or to specimens escaping from fish farms.

***Sarotherodon galilaeus* (Linnaeus, 1758)**

Introduction to Lower Guinea: Five subspecies are recognized by TREWAVAS (1983), two of which occur naturally in Lower Guinea: *S. g. galilaeus* (Linnaeus, 1758) and *S. g. sanagaensis* (Thys van den Audenaerde, 1966).In addition, *Sarotherodon g. galilaeus* was also introduced into Lower Guinea, particularly after 1955, from the Nilo-Sudanian region, Fort-Lamy (12°07' N-15°03' E), Chad, to Yaoundé (3°52' N-11°31' E), Cameroon.

It was also introduced in 1953 from Sudan to the Republic of the Congo.

According to the FAO (2005), the species reproduces and has established itself in the Republic of the Congo. Teugels *et al.* (1991) in Daget *et al.* (1991) report the presence of *S. galilaeus* in the Kouilou basin in the Republic of the Congo and mention the possibility of an introduction.

Sarotherodon g. sanagaensis was transferred in Lower Guinea from the Lom River (5°20'–6°40' N – 13°30'–15°00' E), part of the Sanaga basin, to the vicinity of Betare-Oya (5°36' N–14°05' E), Cameroon.

According to Trewavas (1983), a [third]subspecies, *S. g. boulengeri* (Pellegrin, 1903), can be found in the lagoons of the Lower Ogooué, Gabon. Since the species is not native to this area, this must be the result of an introduction.

The presence of *S. galilaeus* in Lower Guinea is confirmed by museum specimens from Gabon and the Republic of the Congo.

***Tilapia rendalli* (Boulenger, 1897)**

Description: Typically high-bodied, with a body height of 42.2–49.4% of total body length (TBL). Head length 31.1–37.5% of TBL; head profile convex, though in larger specimens it is sometimes concave due to jaw development. Protruding mouth with prominent bicuspid teeth that are relatively short, broad, and very stout. Lower pharyngeal bone approximately as long as it is wide, its dentate portion longer than the anterior portion. 7–10 branchiospines on the lower part of the first gill arch. Dorsal fin with 14–16 spines and 12–13 soft rays; anal fin with 3 spines and 9–10 soft rays. Cycloid scales, 28–32 along the lateral line; cheeks with 3–5 rows of scales. Juveniles are recognizable by their rounded head and beak-shaped mouth.

Maximum size: approximately 400 mm LT and a weight of 2 kg. Coloration: mature specimens generally gray, dorsal side dark olive-green to brown, ventral side with a bright red throat and chest. Body with 5–7 broad dark olive vertical bars. Dorsal surface often dotted with blue scales with a black- d spot at their base. Soft parts of the dorsal and anal fins and the lower half of the caudal fin vary from yellow to red. Tilapia spot always clearly present. Juveniles are recognizable by a few broad body bands, unpigmented pelvic fins, and a bar pattern on the soft dorsal fin.

Natural distribution: Known from Katanga and the Lualaba River as far as Kisangani (Stanleyville)-Isangi, D.R. Congo. Also known from Lakes Tanganyika and Malawi, the Cunene, Okavango, and Zambezi basins, and coastal rivers and lakes in East Africa from the Zambezi Delta to the Pongolo Basin in the south. Introduction to Lower Guinea: species introduced as *Tilapia melanopleura* for aquaculture purposes in 1949, from Yangambi (0°47' N-24°28' E), DRC, to Yaoundé (3°52' N-11°31' E), Cameroon. According to Thys van den Audenaerde (1966), who observed a few specimens of *T. rendalli* at the Melen fish farm (3°52' N-11°29' E) near Yaoundé, the specimens from Yangambi originated from Katanga.

Its use in fish farming in Cameroon has been discontinued. According to the FAO (2005), the species has not established itself in the region. However, according to Moreau *et al.* (1988), the species has acclimatized in the Djerem (5°20' N-13°24' E) and Noun (4°55' N-11°06' E) rivers, in the Sanaga basin. Its presence in Cameroon is confirmed by museum specimens.

It was also introduced for fish farming purposes in 1953 from the Democratic Republic of the Congo to the Djoumouna fish farm (4°23' S–15°10' E) in the Republic of the Congo, in the Congo Basin. According to Moreau *et al., op. cit.,* its use in aquaculture has been discontinued, but the species has likely become established in the region. The introduction of *T. rendalli* into the Lower Guinea region of the Republic of the Congo is confirmed by a museum specimen from a pond at the Dimonika station (4°14' S-12°25' E). The species was also introduced as *T. melanopleura* around 1950 from Katanga but originating from fishing sites in Kinshasa (Léopoldville), DR Congo, near Libreville, Makokou, Lebamba, Franceville, and possibly other locations in Gabon. Its presence in Gabon is confirmed by museum specimens from the Ogooué Basin.

***Tilapia zillii* (Gervais, 1848)**

Description: Body fairly high, height approximately 55% of total length (TL). Head length 31.9–34.1% of TL; head profile straight, slightly oblique. Outer teeth of the jaw bicuspid, not spatulate. Lower pharyngeal bone as long as it is wide, its dentate portion longer than the anterior portion; posterior pharyngeal teeth tricuspid; median pharyngeal teeth not enlarged.

8–12 gill spines on the lower part of the first gill arch. Dorsal fin with 13–16 spines and 10–14 soft rays; anal fin with 3 spines and 8–10 soft rays. Cycloid scales, 28–31 along the lateral line; cheeks with 3–4 rows of scales. Maximum size: approximately 400 mm LS. Coloration: background color in living specimens olive to brownish, with iridescent blue reflections on the scales. Back and sides with 7–10 darker transverse bars and a dark longitudinal band at the level of the lower lateral line; a second longitudinal band is sometimes present along the upper lateral line. Light green lips. Dorsal, anal, and caudal fins olive to brown with yellow spots; dorsal and anal fins often surrounded by an orange band.

The caudal fin has an irregular pattern of darker lines. A large tilapia-like spot extends from the last spiny to the^{4th}soft ray and is always bordered by a yellow band.

Very distinct opercular spot. Iridescent green spots on the head in sexually mature individuals. In juveniles, the black spots on the dorsal and anal fins are orange, and the caudal fin appears vertically striated.

Natural distribution: Known from southern Morocco, the Sahara, and various basins in West Africa such as the Senegal, Ogun, Sassandra, Volta, Niger, and Bénoué rivers. Also found in the basins of Chad, Ubangi-Uélé-Ituri, Lakes Albert and Turkana, and the Nile and Jordan river systems.

Introduction to Lower Guinea: species introduced in 1949, as *Tilapia melanopleura*, from the Yangambi fish farm (0°47' N-24°28' E), D.R. Congo, to Yaoundé (3°52' N-11°31' E), Cameroon. The name *T.* melanopleura has also been used for *T. rendalli*. Apparently, a hybrid between this population of *T. rendalli* originating from Katanga and *T. zillii* was created at Yangambi. The latter, although not very common in the Kibali-Ituri forest biotope (Thys van den Audenaerde, 1996), had gradually replaced *T. rendalli* in the Yangambi ponds.

This hybrid was transported to Cameroon, and consequently, the specimens were at least partly *T. zillii*. The presence of *T. zillii* is confirmed by several museum specimens.

3.2.1.13. Family Clupeidae

The Clupeidae family is best known for its marine species (sardines and herring). It includes several genera and species that have adapted to freshwater and brackish water, as well as others that enter lagoons, such as the species of the Pellonumac subfamily (Gourène *et al.,* 1994). Like all Clupeomorpha, Clupeidae are characterized by an otophyseal connection between the swim bladder and the inner ear, which significantly enhances auditory perception. Most Clupeidae possess several ventral or pelvic scutes that allow for rapid identification of the family. They are overwhelmingly pelagic and gregarious marine teleosts. However, they have colonized certain inland waters where the subfamily Pellonulinae has diversified extensively.

Key to the genera

Premaxillary and dental teeth lacking larger canines*Microthrissa*

Species key

Sparse presence of ventral prepelvic scutes located anterior to the point of insertion of the pectoral fin base; relatively fine premaxillary dentition directed toward the interior of the mouth *M. royauxi*

***Microthrissa royauxi* Boulenger, 1902**

Observed characteristics

Presence of 15 to 17 gill rakers, keeled abdominal scales. The abdominal keel is relatively weak, especially anterior to the ventral scales, a character that brings this specimen closer to *Stolothrissa lembessensis* Fowler (1936). Maximum size 23.5% LT. The entire body is silvery and blue-green.

Description

Pelagic, fusiform, and compressed teleosts with short fins lacking spiny rays; the unpaired fins are short, the anal fin slightly longer than the dorsal fin, and the caudal fin forked. The paired fins are widely spaced. The head is unscaled; the mouth has a non-protractile upper jaw; dentition is generally weak, consistent with a microphagous diet. The lateral line usually lacks pores on the trunk (with the exception of the Denticipitidae); these are small fish.

No pharyngeal teeth; swim bladder is a physostome. Clupeidae are plankton-feeding fish with highly developed gill rakers; they play a major role in the food chain.

Geographic range

According to Gourène and Teugels (1990); Gourène and Teugels (1991a); and Teugels (2007b), the species is widespread in the Congo Basin.

2.3.5. Family Hepsetidae

2.3.5.1. *Hepsetus odoe* (Bloch, 1794)

Description: The body height is 4.5 to 5.8 times the standard length, and the head length is 3 to 3.7 times the standard length. The head, with a flat upper profile, is 2.5 to 3 times as long as it is wide. The maxilla extends backward beyond the posterior edge of the eye. The eye's diameter is 6 to 8 times the head length. There are 9 to 15 gill spines on the gill rake and 3 to 12 on the gill spine.

There are 45 to 49 vertebrae. The scalar formula is: 8.5–10.5/49–60/7.5–9.5.

The dorsal fin is located in the space between the ventral fins and the anal fin.

Maximum size: 445 mm LS with a weight of 822 g in the Ogun River (Moriarty, 1983). A specimen measuring 510 mm LS has been reported in Senegal, though no further details are available.

Coloration: The back is brownish and the belly is white. The scales on the flanks are edged with dark coloration, forming a dark network with light-colored meshes. On the flanks, there are also dark spots or vertical bands, particularly visible in adults.

The upper part of the eye is marked with reddish-orange.

Generally, light and dark bands radiate from the eye toward the rear. The adipose fin is black at least in its distal portion. The fins are pinkish or grayish. In adults, the odd-numbered fins are additionally marked with small black spots.

Distribution: In Lower Guinea, *H. odoe* occurs in most of the region's river basins: Cross, Wouri, Sanaga, Nyong, Kienke (= Kribi), Ntem, Benito (= Mbini, = Woleu), Como, Ogooué, Nyanga, Kouilou, Loeme, and Chiloango. Elsewhere, it is found in numerous basins of West Africa as far as the Zambezi, in the Congo Basin (including the Dja, Sangha, Lake Tumba, and Lake Moero), and in the Cunene and Okavango.

2.4. Taxonomy of the most commonly caught and sold fish species in Kindu

In this section of the study, we will present, for each species, the taxonomic classification to facilitate a clear understanding of the species, as well as the measurements taken

2.4.1. Systematic position of *Chrysichthys wagenaari*

Using the ichthyological classification for teleosts, Nelson (1994) describes the systematic position of *Chrysichthys wagenaari* (Boulenger, 1899) as follows:

- Kingdom Animalia
- Phylum Chordata
- Subphylum .... Chordata
- Superclass..Osteichthyans
- Class Actinopterygi

-Subclass Neopterygian

Infraclass Teleostei

-Superorder Ostariophysii

-Order ....Siluriformes

- Suborder Siluroidei

- Family Bagridae

- Genus *Chrysichthys*

- Species *C. wagenaari*

2.4.1.1. Morphological characteristics

The species *Chrysichthys* wagenaari has a very flattened head, as long as it is wide, rough on top, with an occipital process [5]barely longer than wide and in contact with the interneural plate, and a broadly truncated snout that extends only slightly beyond the mouth. The nasal barbels are slightly longer; the outer mandibular barbels are longer than the inner ones—measuring half or two-thirds of the head length. The pectoral spine measures 3/5 of the head length, slightly serrated on the outer edge, very strongly so on the inner edge; the caudal spine is deeply notched with pointed lobes. The caudal spine is longer than the pectoral spine, brown above, white below, and silvery along the edges. Sometimes two blackish transverse bands behind the head, the second crossing the base of the dorsal fin at its origin; brown fin (Poll, 1957).

Naked fish, moderately elongated, with a radiated dorsal fin and two pectoral fins, each bearing a spine. Adipose fin present and small. Anal fins short with 8 branched rays (Devos, 1986, cited by Laleye, 1995). Related to the Ariidae, but the gill membrane is more or less arched in the middle, free, and briefly fused to the isthmus.

In this family, the Clarotoidae, the genus *Chrysichthys*, which grows larger and heavier, has, according to Risch (1986): 4 pairs of barbels.

- A dorsal fin with 6 (rarely 5 or 7) soft rays, preceded by a very short spine and a strongly developed spine that is slightly denticulate along its posterior margin;

- A medium-sized or small adipose fin (the base being smaller than the width of the head), and never ossified;

- A pair of pelvic fins, located approximately in the middle of the body, with a single ray and five branched rays. A medium-sized anal fin with 3–6 simple rays and 6–12 branched rays; a well-forked caudal fin. The eyes, with free rims, are lateral and large. The body is moderately elongated, 4–6 times as long as it is high.

2.4.1.2. Ecological considerations

As warm-water fish, *Chrysichthys* species inhabit freshwater and brackish waters (Daget and Iltis, 1965; Risch, 1992). They are found in both large lakes and rivers (Lowe Mc Connell, 1975) and in small lakes and rivers in Africa (Ikusemiju, 1976; Albaret, 1982). They rank among the heaviest freshwater fish in Africa (Risch, 1986).

Their requirements regarding salinity and the characteristics of their migrations between lagoons and rivers have not yet been the subject of specific studies (Laleye, 1995).

2.4.1.3. Geographic distribution

According to Jayaram (1966), the range of the genus *Chrysichthys* extends from the Gambia River basin in the west to Lake Tanganyika in the east. The recent systematic revision by Risch (1986) indicates a more extensive distribution. Indeed, *Chrysichthys* have been found in the Limpopo River in southeastern Africa and throughout the Nile Basin. The new classification lists ten subgenera, of which only the genera *Chrysichthys, Melanodactytus,* and *Rheoglanis* are represented in Africa (Risch, 1992). The others can be found in both Asia and tropical Africa (Poll, 1957).

2.4.1.4. Feeding

Clarotoids can be divided into two main groups: the *Clarotes* and *Bagrus* on one hand, and the *Auchenoglanis* and *Chrysichthys* on the other. *Chrysichthys* and *Auchenoglanis* are primarily consumers of insect larvae, particularly *Chironomids.* They are benthic (Lowe Mc Connell, 1973) and omnivorous fish that feed on bivalves, crabs, copepods, fish eggs, and fry (Fagade and Olaniyan, 1973), cited by Laleye (1995). Plants and organic debris have also been found in their stomachs (Chauvet, 1973; Lowe Mc Connell, 1975).

Supplementary food may consist of small mollusks, fry, swimming hemipterans, and zooplankton, especially in juveniles (Lauzane, 1988 in Lévêque *et al.*, 1988).

2.4.2. Systematic Considerations of *G. numenius*

The taxonomy of *G. numenius* is as follows:

Phylum Chordata

Subphylum Vertebrata

Class Teleostomi

Subclass Actinopterygii

Order Clupeiformes

Family Mormyridae

Genus Gnathonemus BLGR

Species: *Gnathonemus numenius* BLGR,

2.4.2.1. Morphological and ecological characteristics

G. numenius is a highly compressed and somewhat elongated fish, with a relatively smaller posterior body than in other fish and a small caudal fin. The other vertical fins—dorsal and anal—are well-defined. The paired fins are shortened, and the soft fins lack spines. The scales are cycloid and small. Like all Mormyridae, the posterior section is notably pedunculated and contains an electric organ of muscular origin (evident in only a few cases), of low power and whose intensity is not detectable without a highly sensitive detection device.

Recent research conducted in the USA has shown that this electric organ is related to the mormyromasters and that together they form a very unique radar system. The mouth is bordered by very small premaxillae and maxillae, supported by a tubiform snout acquired through polygenic evolution. The teeth on the jaws are bicuspid or conical and very often feature the parasphenoid and the tongue. No pharyngeal teeth; a simple, elongated swim bladder connected to the inner ear and linked to the esophagus by a canal. This fish breathes solely through gills; the gill slits are well-developed. It is a fish that sometimes builds a nest in vegetation and prefers euphotic and aphotic benthic zones, more or less muddy (Poll, 1957).

2.6.4. Systematic Considerations of *Labeo lineatus*

According to Poll and Gosse (1995), this species belongs taxonomically to:

Phylum: Chordata

Subphylum: Vertebrata

Class: Ostariophysi

Order: Cypriniformes

Family: Cyprinidae

Genus: Labeo

Species: *Labeo lineatus* (Boulenger, 1907)

2.6.4.1. Morphological description and geographic distribution

The body is generally fusiform and resembles that of the Characidae but is always moderately compressed and never bears an adipose fin.

The vertical fins are short; the dorsal and anal fins are superimposed, with few exceptions, and the caudal peduncle is free. The fin rays are all soft, except occasionally for the dorsal fin, which may have an ossified ray at the front. The caudal fin is generally notched. Paired fins; pectoral and pelvic fins are widely spaced. The pectoral fins are located at a low insertion.

Cycloid scales are fairly large, lateral, and complete medially; rarely reduced to the anterior scales only, and sometimes absent. The head is bony, protractile, terminal, and either superior or inferior, bordered only by the premaxillae, and may or may not have one or two pairs of labial barbels on the upper lip. In the case of a terminal or inferior mouth, a sucker may be narrow and covered by a horny sheath.

The buccal teeth are always absent. In contrast, the pharyngeal bones bear teeth in one or more rows that play an important role in mastication.

Known in both Lower Congo and the Central Basin and Katanga, this species has already been caught in Leopoldville (now Kinshasa).

Pellegrin (1932) also had the opportunity to mention a giant specimen of *Labeo lineatus* from the "Brazzaville region." This specimen measured 700 millimeters.

2.6.4.2. Ecology of the species

Labeo species are either found in fast-flowing waters with rocky bottoms or in calmer waters with muddy, especially sandy, bottoms. They all feed on microorganisms in the bottom vegetation. Their diet consists mainly of various algae, including many diatoms, but also plant detritus, microcrustaceans (copepods, cladocera, and ostracods), and rotifers.

2.6.5. Systematic considerations *of Clarias gariepinus*

The systematic classification of *Clarias gariepinus* is as follows:

Phylum Chordata

- Subphylum Craniata

Superclass Gnathostomata

Class Osteichthyes

Subclass Actinopterygii

Infraclass Teleostei

Order Siluriformes

Suborder Siluroidei

FamilyClariidae

Genus *Clarias*

Species *C. gariepinus, Burchell (year)*

C. gariepinus has an omnivorous diet with a carnivorous tendency. The larvae feed on zooplankton, while juveniles primarily consume Chironomid larvae and other aquatic insects. As adults, they feed primarily on aquatic insects, fish, mollusks, and plant debris, etc. (Micha, 1973; Bruton 1979; Spataru *et al.*, 1987).

1.1.1. 1. Biology and Ecology

C. gariepinus exhibits seasonal gonadal maturation linked to rainfall (Halim and Guma'a, 1989; Dadzie and Ochieng-Orach, 1989; Merron and Mann, 1995; Yalçin *et al.*, 2001). The process of gonadal maturation is influenced by annual changes in water temperature and photoperiod, and spawning is triggered by a rise in water level (De Graaf *et al.*, 1995). The breeding season varies by river when water temperature ranges between 21 and 30°C (Yalçin *et al.*, 2001). Adults of *C. gariepinus* migrate to floodplains during the breeding season (Micha, 1975; Douglas, 1979; Dadebo, 2000) as soon as the water current permits. Baijot *et al.* (1994) report that *C. gariepinus* takes refuge in the grass beds that form along the edges of watercourses in anticipation of rising water levels. The table below highlights reproduction in various habitats.

The size at first sexual maturity observed in *C. gariepinus* in the wild varies by watercourse, ranging from 24.7 cm to 75 cm in males and from 23.3 cm to 75 cm in females (Yalçin *et al.*, 2001). Intraspecific variation in fecundity, attributed to variation in the size of females constituting the spawning stock in *C. gariepinus* populations, has been observed (Jobling, 1995). This reproductive parameter ranges from 3,000 to 336,157 eggs per female

2.3.4. Fish measurements

The weight-size relationship requires, above all, a better choice of measurement that best represents length. On this point, researchers' opinions are divided: some have opted for total length (Hile, 1948; Micha, 1973; Cuinat, 1975), while others (Daget, 1954 and 1962; Blache, 1964; Teugels, 1986; and Nyongombe, 1993b) have adopted standard length.

Total length here corresponds to the maximum straight-line distance between the tip of the snout and the tip of the longest ray of the caudal fin. It is easy to measure but unreliable due to various accidents that may occur during the fish's life, or during capture or preservation in formalin (Mutambwe, 1984; Teugels, 1986).

As for standard length (SL), it corresponds to the length between the tip of the snout and the base of the caudal fin. This length provides a fairly accurate indication of the fish's actual size. Weight here refers to the total weight of the fish as recorded at the fishing site.

2.3.5. Analysis of stomach contents, sexual maturity size, and sexual dimorphism

A binocular microscope (Eleitz Wetzlar, 10x magnification), a standard microscope (Reichert Austria, 100x magnification), and a dissection kit were used to examine the stomach or intestinal contents of the fish.

The analysis of fish digestive tract contents has already been the subject of a relatively extensive body of literature. Several authors, including Hulot (1950), Micha (1973), and Nyongombe (1993a and b), have studied fish diets using one or the other of the following methods:

A. Volumetric and gravimetric methods

These methods involve measuring the volume or dry weight at temperatures ranging from 60° to 105°C, or the wet weight of each prey category. These methods allow for determining the total amount of food present in the digestive tract and provide a good estimate of the relative importance of the different prey categories. Unfortunately, they are very time-consuming.

B. Occurrence

With this method, the number of stomachs or intestines in which a given prey item is found is recorded and expressed as a percentage of the total number of stomachs or intestines analyzed. This is a quick, simple, and easy method that requires minimal equipment for the species in question. However, it has the disadvantage of not providing an exact idea of the total amount of food ingested by the fish or the relative importance of each prey item (Nyongombe, 1993ab).

C. Dominance method

In this method, the number of stomachs in which a given prey item visibly dominates all others in terms of volume or weight is counted, and this is expressed as a percentage of the total number of stomachs analyzed.

Some authors find this method to be relatively subjective because the dominant prey is sometimes difficult to determine.

In practice, however, the dominance method is more easily reproducible than the "point" method, in which a certain number of points is arbitrarily assigned to each prey item based on its abundance and size, with the total points obtained expressed as a percentage of the total points allocated to all prey items.

Given the objective pursued and the materials available to us, we chose the occurrence method. This allowed us to define the range of prey on which each captured specimen feeds and to determine its preference (Nyongombe, *op. cit.*).

Regarding the sex ratio, this parameter was determined by establishing the ratio between the number of male and female fish within the same species.

Sexual dimorphism was determined by direct examination of the urogenital tract following abdominal compression of the fish. If the rectum (anus) retracted into the fish's body like a miniature anal prolapse, the fish was female.

Conversely, if the rectum (anus) protrudes from the fish's body like a miniature anal prolapse, the fish is a male. This dimorphism was also observed after eviscerating the abdominal contents with a scalpel to locate the gonads.

2.3.6. Marketing Process

Information regarding marketing was obtained using a segmented survey method.

2.3.7. Specimen Preservation

The captured specimens were taken to the laboratory of the Faculty of Agricultural Sciences for preservation in 4% formalin (for fixation) in jars.

2.3.8. Statistical analysis

The various data were analyzed using Excel 2007 to calculate arithmetic means, percentages, and frequencies.

Relative abundance is calculated using the formula below

- Relative abundance: R.A. = (n/N) x 100

Where

- RA = relative abundance of the species under consideration;
- n = number of individuals of a species;
- N = total number of individuals of all species combined (DAJOZ, 1975).

2.3.9. Challenges Encountered

We must admit that not everything went smoothly along the way. We certainly encountered many difficulties throughout our study, notably:

1. the lack of actual data on the fish inventory in this specific section of the river, particularly in the Kindu

area;

2. the lack of a laboratory in the region, which could have helped us analyze other physicochemical parameters not included in this study.

Chapter Three: Presentation, Interpretation, and Discussion of Results

In this chapter, we will first present, interpret, and discuss the results regarding the observed physicochemical parameters, followed by the results on the biodiversity of fish caught in this part of the Congo River basin; we will then present the fishing gear used and seek to gain insight into fish sales activities in the markets of Kindu.

3.1. Physicochemical analyses of the water

3.1.1. Water temperature

In the study area, we observed that the recorded annual average temperature is 26.5°C (Figure 6).

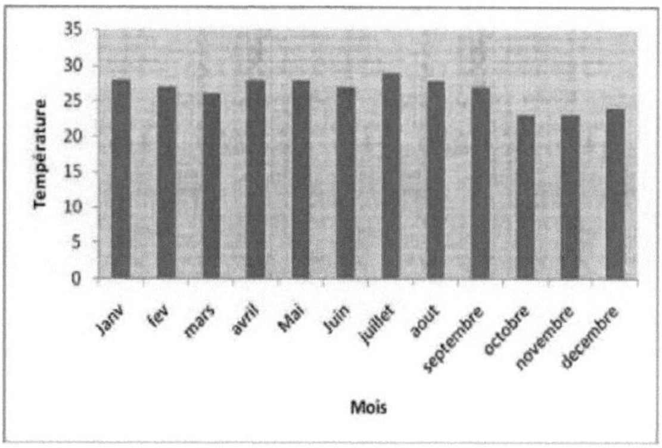

Figure 6: Annual temperature trend.

The temperature data presented in the figure above show that the average temperature observed in the Congo River is 26.5°C. The minimum and maximum temperatures range from 23.0°C to 29.0°C. This indicates that in this section of the Congo River, the water is suitable for fish reproduction, growth, and development.

3.1.2. Water pH

The results presented in Figure 7 below reveal that the water of the Congo River in this section of Kindu fluctuates around neutrality, i.e., (pH = 6.88 to 7.07). This result is consistent with the literature, which indicates that the pH in natural waters ranges between 6 and 8.5 (Chapman *et al.*, 1996). This situation may be explained by significant human activity in the studied area.

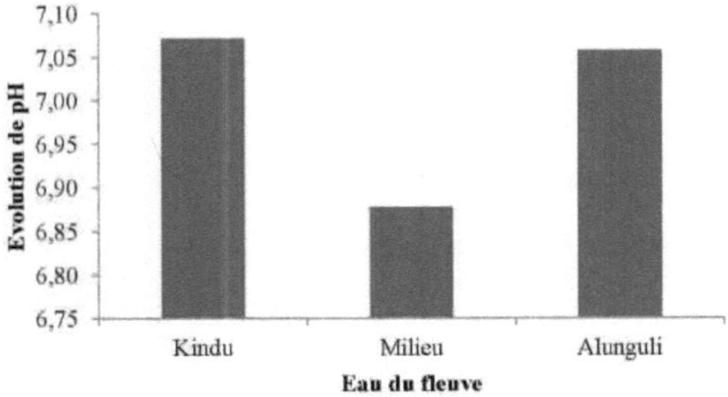

Figure 7: Annual hydrogen ion concentration

3.1.3. Dissolved oxygen (O_2) in mg/L

Dissolved oxygen measures the concentration of dissolved oxygen in water (Rodier, 1984). It plays a role in most chemical and biological processes in aquatic environments. Table 4 shows the measured oxygen levels during our study.

Table 4: Evolution of oxygen levels.

Parameter	Minimum	Maximum	Average
Oxygen (in mg/L)	6.3	8.0	7.2

The results for dissolved oxygen in the water, as presented in Table 4 above, show that the average oxygen level observed in the river water is approximately 7.2 mg/L. The minimum and maximum oxygen levels range from 6.3 to 8.0 mg/L, indicating good oxygenation at the study site and favorable conditions for the development, growth, and reproduction of fish.

3.1.4. Transparency

Table 5 shows that the waters of the Congo River in the Kindu section have a transparency of 47.8 cm toward the city (left bank); on the right bank (Alunguli), it is 45.6 cm; and in the middle of the river, there is an average of 59.2 cm.The transparency values found in our study do not exceed the 30–60 cm range reported by Mulangu (2012).

Table 5: River water transparency (in cm)

Parameter	Left bank (Kindu)	Middle	Right Bank (Alunguli)
Water transparency (cm)	47.8 cm	59.2 cm	45.6 cm

These results are similar to those found by Mulangu (2012) in the Lubilanji River in Eastern Kasai.

3.2. Fishing techniques

Local fishing in Kindu is practiced by several tribes, primarily the Genia, Lokele, Kumu, and Topoké, who have come from Kisangani by traveling up the Congo River.

The various gear types used include: gillnets, trammel nets, fine-mesh nets, and traps. In total, we surveyed sixty fishermen using various types of gear to catch fish. A total of 2,318 fish were caught, as shown in the table below.

Table 6: Number of fish caught by each gear type

Type of gear	Number of fish	Percentage
Set net	1,290	55.65
Gillnet	604	26.05
Gill net	305	13.15
Pot	119	5.13
Grand total	2,318	100%

Analysis of this table shows that the most commonly used fishing gear is the gillnet. Of the total number of fish caught, this gear accounted for 55.65%. The gillnet comes in second place, accounting for 26.05% of the fish caught, followed by the trammel net at 13.15%. The least-used gear is the trap, which was used to catch only 5.13% of the fish.

Apart from the various fishing techniques mentioned above, there is another fishing technique that is not easily explo ly and is carried out clandestinely because it is prohibited by current regulations. This involves the use of plants containing dangerous phytotoxins such as *Millettia macroura* Harms, *Euphorbia sp. (Potenge*), and *Tephrosia vogelii.*

Photo 8: *Millettia macroura* Harms, *Euphorbia sp. (Potenge*), and *Tephrosia vogelii*

3.3. Caught fish

Table 7 shows the number of fish caught.

Table 7: Number of Fish Caught.

Species	Number	(%)
Clarias gariepinus	200	8.62
Clarias angolensis	140	6.03
Clarias buthypogon	152	6.55
Clarias camerunensis	19	0.81
Chrysichthys nigrodigitatus	32	1.38
Chrysichthys auratus	42	1.81
Chrysichthys wagenaari	60	2.58

Oreochromis macrochir	49	2.11
Oreochromis mossambicus	52	2.24
Tilapia zillii	142	6.12
Oreochromis niloticus	152	6.55
Sarotherodon melanotheron	100	4.31
Sarotherodon galilaeus	110	4.74
Gnathonemus numenius	280	12.07
Petrocephalus ballayi	20	0.86
Petrocephalus microphtalmus	7	0.30
Mormyrops nigrican	13	0.56
Distichodus rostratus	7	0.30
Distichodus fasciatus	11	0.47
Hepsetu odoe	6	0.25
Ctenpoma petrecii	10	0.43
Xenomystus nigri	6	0.25
Papyrocranys afer	11	0.47
Barbus callipterus	12	0.51
Barbus leonensis	8	0.34
Labeo lineatus	118	5.09
Heterotus nilocutus	222	9.57
Schilbe intermedius	12	0.51
Schilbe grefeli	15	0.64
Malapterurus electricus	68	2.93
Malapterurus lacepede	42	1.81
Citharinuus gibbosus	106	4.57
Citharinus congicus	62	2.67
Synodontis batesii	32	1.38
Total	2,318	100.00

The results presented in this table sufficiently demonstrate that the species *Gnathonemus numenius* (12.07%), *Heterotus niloticus* (9.57%), *Clarias gariepinus* (8.62%), *Oreochromis niloticus*, and *Clarias bithupogon*, have respectively (6.55%),*Tilapia zillii* (6.12%), *Clarias angolensis* (6.03%), *Labeo lineatus* (5.09%), *Sarotherodon galilaeus* (4.74%), *Citharinus gibbosus* (4.57%), *Sarotherodon melanotheron* (4.31%), *Malapterurus electricus* (2.93%), *C. congicus* (2.67%), *Chrysichthys wagenaari* (2.58%), and *Oreochromis mossambicus* (2.24%) are the most commonly caught in this section of the Congo River and sold in the various markets of Kindu. This situation can be explained by their adaptation to the environmental conditions and the artisanal fishing practices carried out there.

In addition, various other species are also caught, such as *Oreochromis macrochir* (2.11%), *Chrysichthys* auratus and *Malapterurus lacepède* (1.81% each), *Chrysichthys nigrodigitatus* and *Synodontis batesii* (1.38% each), *Petrocephalus ballayi* (0.86%),*Clarias camerunensis* (0.81%), *Schilbe grefeli* (0.64%), *Mormyrops nigrican (*0.56%), *Schilbe intermedius* and *Barbus callipterus* respectively (0.51%), *Distichodus fasciatus* and *Papyrocranys afer* respectively (0.47%),*Ctenopoma petrecii* (0.43%), *Barbus leonensi* (0.34%), *Distichodus rostratus* and *Petrocepahlus microphtalmus* respectively (0.30%), *Hepsetu odoe* and *Xenomystus nigri* (0.25%).

The most frequently captured families are the Cichlidae with 6 species (17.14%), followed by the Clariidae and Mormyridae families with 4 species (11.43%), then the Clarotoidae and Cyprinidae families with 3 species (8.57%); the families Distichodontidae, Notopteridae, Schilbeidae, Malapteruridae, and Citharinidae are each represented by 2 species (5.71%).

The remaining families are the least frequently caught, each represented by a single species, accounting for 2.86% of the total number of fish species inventoried; these are the families Hepsetidae, Anabantidae,

Arapaimidae, Mochokidae, and Channidae, as shown in Figure 8.

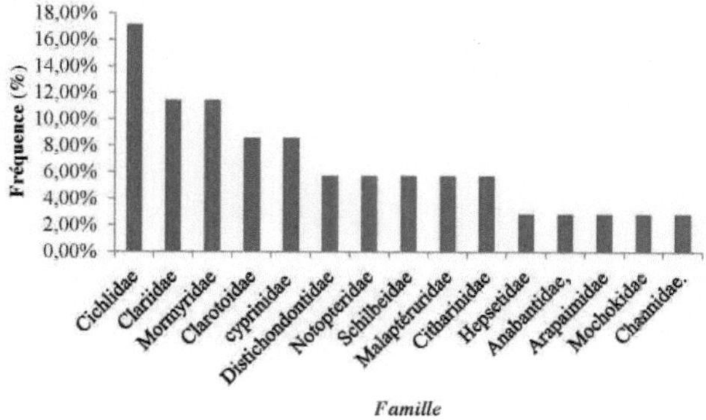

Figure 8: Frequency of Fish Families Caught

Figures 9 through 16 show the relative abundances of species by family.

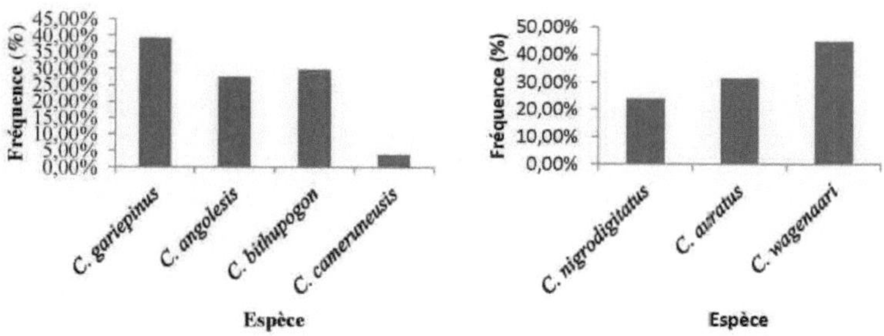

Figure 9: Relative Abundances of Clariidae and Clarotoidae Species

Regarding the Clariidae, we note that *Clarias gariepinus* ranks first. In contrast, for the Chrysichtus family, we note that the species *Chrysichtus wagenaari* ranks first.

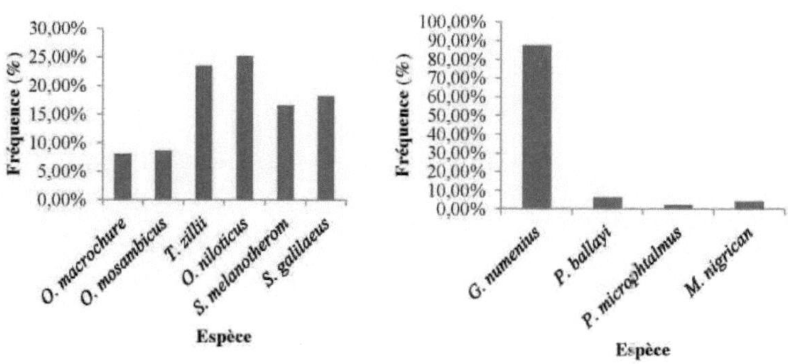

Figure 10: Relative abundance of Cichlidae and Mormyridae.

Regarding relative abundance among Cichlidae, we observe that the species *Oreochromis niloticus* ranks first, followed by *Tilapia zillii.* In contrast, among Mormyridae, the species *Gnatonemus numenius* is the most abundant.

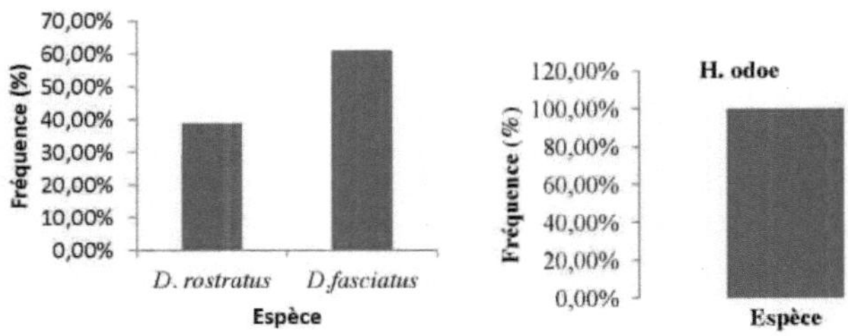

Figure 11: Relative abundance of Dischodontidae and Hepsetidae.

The relative abundance among the Distichodontidae indicates the dominance of the species *Distichodus faciatus*. In contrast, the Hepsetidae family is monospecific.

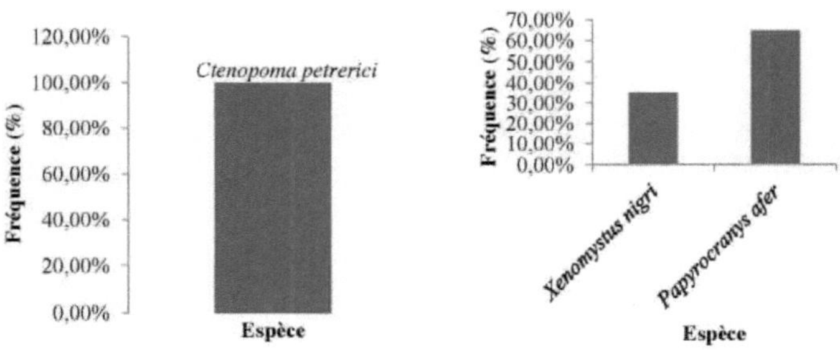

Figure 12: Relative abundances of the Anabantidae and Notopteridae.

The relative abundance within the two families is as follows: among the Anabantidae, there is a single species, *Ctenopoma petrerici.* In contrast, *among the Notopteridae, Papyrocranys* afer dominates.

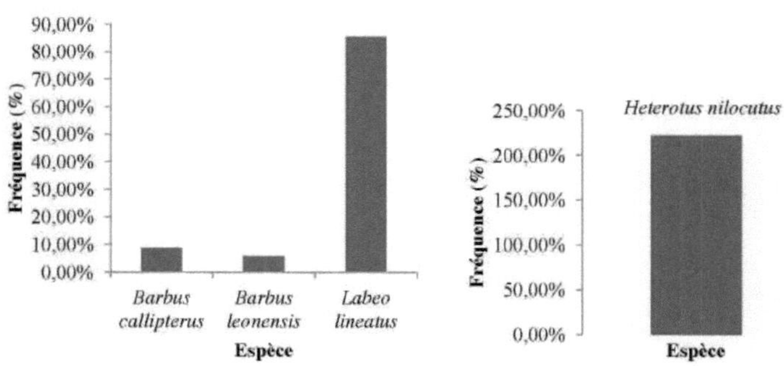

Figure 13: Relative abundances of Cyprinidae and Arapamidae (Osteoglossidae).

Amonġ the Cyprinidae, *Labeo lineatus* dominates with a frequency of over 80%. In contrast, the Arapamidae family has only one *species, Heterotis niloticus.*

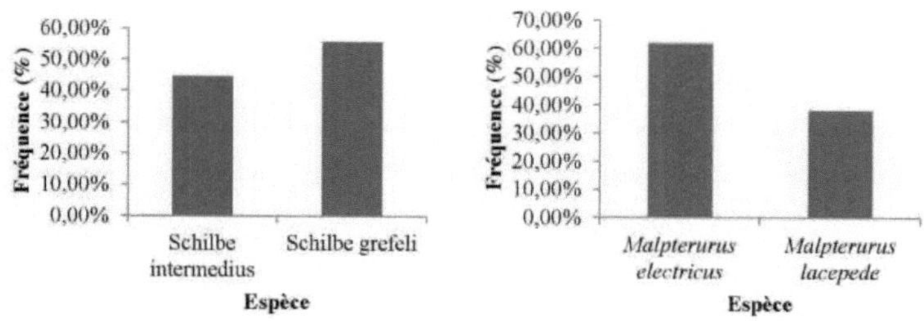

Figure 14 Relative abundances of the Shilbeidae and Malapteruriidae.

Among the Shilbeidae: *Shilbe grefeli* is the most abundant. Among the Malapteruriidae, *Malapterus electricus* is the most abundant.

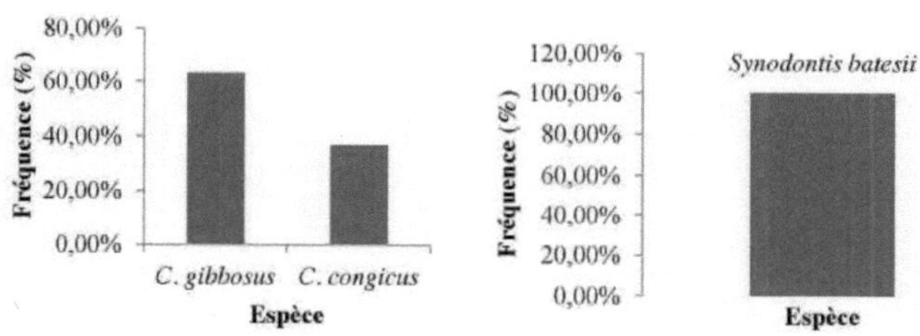

Figure 15 : Relative abundances of the Citharinidae and Mochokidae.

Among the Citharinidae: the species *Citharinus gibbosus* is the most abundant. In contrast, among the Mochokidae, we have only one species, *Synodontis batesii.* The same is true for the Channidae.

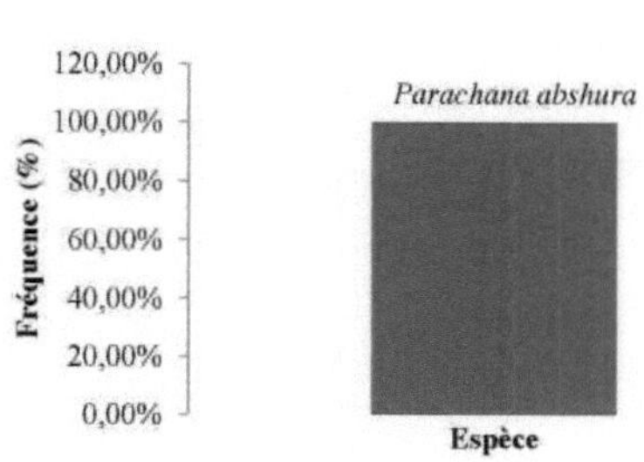

Figure 16 : Relative abundance among the Channidae.

3.4. Fish measurements and sex ratio

3.4.1. Some metric and meristic characteristics of the preferred fish (high economic value)

Information regarding the metric and meristic parameters of the preferred fish species is presented in the tables below.Regarding these parameters, we focused on the four most commonly caught and sold fish species that can increase the stock of fish to be raised in Kindu and thus help reduce malnutrition in the city of Kindu. Several parameters were observed. These include qualitative characteristics, metric characteristics, and meristic characteristics.

Qualitative characteristics: These concern observable features on the fish, such as the type of fins and scales, the shape of the mouth cavity, the presence or absence of barbels and teeth, the coloration of the specimen, the position of the paired or unpaired fins, the presence or absence of dermal openings, the shape of the frontal fontanelle, the shape of the urogenital papilla, the shape of the caudal fin, etc.

Meristic characters: involve counting the number of teeth, lateral lines, gill rakers, pairs of gill arches, fin rays, *serrations* on the pectoral spine, etc. Metric characters: involve measurements (TL, SL, and weight taken from the fish). In this study, these characters were used to accurately identify our fish specimens.

3.4.1.1. *Chrysichthys wagenaari*

Tableau 10 : Overall statistics on the indicators of the male *Chrysichthys wagenaari* species (N=36)

Measurements	Standard length (mm)	Total length (mm)	Weight (g)
Mean	466.7	569.7	556.9
Standard deviation	7.337	6.839	13.401
Minimum	330	420	340
Maximum	650	720	880
Sum	16,800	20,510	20,050

Tableau 11 : Overall statistics on indicators for the male *Chrysichthys wagenaari* (N=24)

Measurements	Standard length (mm)	Total length (mm)	Weight (g)
Mean	412.5	501.3	571.7
Standard deviation	6.117	6.861	13.770
Minimum	320	420	340
Maximum	580	680	980
Total	9,900	12,030	13,720

It follows from these two tables that males are larger (569 mm) than females (501 mm) in total length. The sex ratio (M-F) is skewed in favor of males, at 1.5:1.

Based on the above, we venture to believe that in this fish species, males grow faster than females, all other things being equal. In other words, males reach sexual maturity faster than females. However, this earlier maturity of males compared to females is not surprising, especially since the specialized literature mentions numerous such cases. Beya (1999) noted in *Chrysichthys wagenaari* that males mature earlier (TL = 325.0 mm and SL = 266.50 mm) than females (TL = 345.0 mm and SL = 27.00 mm).

3.4.1.2. *Gnathonemus numenius*

Tableau 12 : Overall statistics for the indicators of the species *Gnathonemus numenius* males (N=151)

Indicators	Total length (mm)	Standard length (mm)	Weight (g)
Mean	103.6	87.79	57.97
Standard deviation	9.047	8.187	23.420
Minimum	91	75	9
Maximum	117	100	121

Sum	15,644	13,257	8,753

Tableau 13 Overall statistics for female *Gnathonemus numenium* (N=130)

Indicators	Total length (mm)	Standard length (mm)	Weight (g)
Mean	132.85	111.18	50.47
Standard deviation	13.203	16.487	20.079
Minimum	120	11	8
Maximum	162	142	117
Sum	17,271	14,454	6561

Both tables indicate that females are larger (132.85 mm) than males (103.60 mm) in total length. This situation is the opposite of the previous case, where males are larger than females. The sex ratio (M - F) is 1.2:1.

3.4.1.3. *Labeo lineatus*

Data on sexual maturity size in *Labeo lineatus* are summarized in Tables 14 and 15.

Table 14: Statistics on overall indicators for male *Labeo lineatus* (N=32)

Measurements	Standard length (mm)	Total length (mm)	Weight (g)
Mean	149.31	166.88	371.9
Standard deviation	32.776	34.757	28.156
Minimum	96	105	80
Maximum	228	253	1260
Sum	4,778	5,340	11,900

Table 15: Statistics on overall indicators for female *Labeo lineatus* (N=28)

Measurements	Standard length (mm)	Total length (mm)	Weight (g)
Mean	157.06	175.50	416.1
Standard deviation	35.261	36.727	19.352
Minimum	99	118	190
Maximum	219	241	890
Sum	2,827	3,159	7,490

The two tables above indicate that females are larger (175.50 mm) than males (166.84 mm) in terms of average total length. The sex ratio (M:F) is 1.1:1. The average weights are 416.1 g and 371.9 g for females and males, respectively.

Data on sexual maturity size in *Clarias gariepinus* are summarized in Tables 16 and 17.

Table 16: Statistics for the entire *Clarias gariepinus* population (males, N=41)

Indicators	Standard length (mm)	Total length (mm)	Weight (g)
Mean	159.9630	179.0000	458.889
Standard deviation	32.67495	32.98718	21.46255
Minimum	76.00	95.00	120.0
Maximum	216.00	$235.00	910.0
Total	4,319.00	4,833.00	12,390.0

Table 17: Statistics for the entire group of female *Clarias gariepinus* (N=35)

Indicators	Standard length (mm)	Total length (mm)	Weight (g)
Mean	147.8077	164.8462	400.4
Standard deviation	33.88335	35.58504	30.057
Minimum	106.00	122.00	90
Maximum	243.00	265.00	1,260
Total	3,843.00	4,286.00	10,410

Both tables indicate that males are larger (179 mm) than females (164.84 mm) in total length. The sex ratio (M:F) is 1.2:1. The average weights are 458 g and 400 g for males and females, respectively.

3.5. Analysis of stomach contents

Analysis of the stomach contents of various preferred fish species shows that these species favor animal-based foods (insects, larvae, fish, scales, and other animal matter), with proportions varying by species. For example, in *Clarias gariepinus,* we identified 71.1% in females versus 66.5% in males.

In Labeo lineatus, we identified 60% in females versus 77.7% in males.

In *Chrysichthys wagenaari,* we observed 63.6% in females and 81.6% in males.

Finally, in *Gnathonemus numenius,* we found 57% in females versus 40% in males.

This analysis shows that our results are similar to those found by other researchers (Nyongmbe, 1993a; Wembodinga, 2014; *Manga et al.*, 2014).

3.6. Fish marketing process in the markets of Kindu

Variations in the selling price of a piece of fish

With regard to this parameter, we surveyed a number of respondents who agreed to complete our questionnaire, and we have presented these results in Table 18.

Table 18: Different retail prices for a piece of fish (in grams)

Species	Weight in grams	Price (FC)	Price per 100 g (FC)
Clarias gariepinus	1,000	13,500	1,800–2,500
Ctarinus gbosus	1,000	12,500	1,500–1,800
Gnahnemus numerous	1,000	9,700	1,400–1,800
Chrysichthys wagenaari	1,000	13,500	1,800–2,000

An analysis of the table above shows that the price of a 100-gram portion of the fish species listed above generally ranges from 1,400 to 2,500 Fc. This situation clearly demonstrates that fish is expensive in the Kindu market.

3.7. Allocation of Income from Fish Sales

Table 19: Allocation of income from fish sales

Allocation of income	Fo	%
Education	4	17.4
Medical care	1	4.4
Food service	7	30.4
Apparel	4	17.4
Purchase of fishing equipment	7	30.4
Total	23	**100%**

An examination of this table shows that of the 23 respondents who agreed to answer our questions—that is, 100%—it appears that income is allocated to food and the purchase of fishing equipment, at 30.4% each, followed by education and clothing, at 17.4% each. Medical care accounts for 4.4%. This shows that fishing enables fishermen to address family issues.

3.8. Methods of preserving unsold catch

The most common preservation method in the community is smoking.

3.9. Number of years spent selling fish

Table 20: Number of years in the fish trade

Number of years	Observed frequencies	%

0 to 10 years	08	26.7
10 to 20 years	13	43.3
20 and under	0.9	30.0
Total	30	100

This table shows that 43.3% of fish sellers have 10 to 20 years of experience selling fresh fish at the Kindu Central Market, while 30% have 20 or more years of experience, and the remainder accounts for 26.7%.

3.10. Additional Activities

When asked whether or not the respondents engage in activities other than fishing, the responses to this question are recorded in Table 21.

Table 21: Additional Activities

Response	Fo	%
No	28	93.33
Yes	02	6.66
Total	30	100

This table shows that 93.3% of female fish sellers do not have any other occupation besides fishing, while 6.6% do.

3.11. Age group of fish sellers

The fishermen were divided into three age groups, as shown in the table below.

Table 22: Age Groups of Fish Sellers

Age group	No.	%
1st Age: ≤30 years	10	33.3
2nd Age: 30 to 45 years	13	43.3
3rd Age: ≥ 45 years	0.7	23.3
Total	30	100

Looking at this table, we see that 43.3% of the vendors fall into the second age group according to the breakdown in our table, while 33.3% are in the first age group and 23.3% are in the third age group. This activity is carried out by heads of households who are still vigorous.

3.12. Time of Fish Catch

The fishermen's opinions regarding the time of fish catch are summarized in the table below.

Table 23: Time of fish catch.

Time of fish catch	Fo	%
Night	27	90
Day	3	10
Total	30	100

A review of this table clearly shows that out of 30 respondents, or 100%, 27 respondents, or 90% , catch their fish at night, while 3 respondents, or 10%, state that they catch their fish during the day.

Conclusion and Recommendations

At the conclusion of this study on the biology and ecology of fish collected from the Congo River and sold in Kindu (Maniema /DR Congo), the following conclusions were drawn:

Regarding physical and chemical parameters:

1. It is observed that the water of the Congo River in this section of Kindu fluctuates around neutrality (pH = 6.88 to 7.07);

2. the average temperature observed is 26.5°C, with minimum and maximum values ranging from 23.0°C to 29.0°C;

3. The dissolved oxygen content averages 7.4 mg/L, with minimum and maximum values ranging from 6.3 to 8.0 mg/L;

4. the average transparency of 47.8 cm was observed on the left bank; that of the right bank is 45.6 cm, while in the middle of the river we recorded an average of 59.2 cm.

Regarding the families and species captured, we observed the following:

- the most dominant families are the Cichlidae with 6 species (17.14%), followed by the Clariidae and Mormyridae families with 4 species (11.43%); next come the families Clarotoidae and Cyprinidae with 3 species (8.57%); the families Distichodontidae, Notopteridae, Schilbeidae, Malapteruridae, and Citharinidae are represented by 2 species each (5.71%);

The remaining, less dominant families are each represented by a single species, accounting for 2.86% of the total number of fish species inventoried; these are the families Hepsetidae, Anabantidae, Arapaimidae, Mochokidae, and Channidae.

Regarding the numerical data on fish caught, we observe significant variability depending on the species.

- The results, as presented, provide sufficient evidence that the species *Gnathonemus numenius* (12.07%), *Heterotus nilocutus* (9.57%), *Clarias gariepinus* (8.62%), *Oreochromis niloticus* and *Clarias bithupogon* (6.55% each), *Tilapia zillii* (6.12%), *Clarias angolensis* (6.03%), *Labeo lineatus* (5.09%), *Sarotherodon galilaeus* (4.74%), *Citharinus gibbosus* (4.57%), *Sarotherodon melanotheron* (4.31%), *Malapterurus electricus* (2.93%), *C. congicus* (2.67%), *Chrysichthys wagenaari* (2.58%), and *Oreochromis mossambicus* (2.24%) are the most commonly caught in this section of the Congo River and sold in various markets in Kindu. In addition, other species are also caught, such *as Oreochromis macrochir* (2.11%), *Chrysichthys auratus* and *Malapterurus lacepède* (1.81% each), *Chrysichthys nigrodigitatus* and *Synodontis batesii* (1.38% each), *Petrocepahlus ballayi* (0.86%), *Clarias camerunensis* (0.81%), *Schilbe grefeli* (0.64%), *Mormyrops nigrican* (0.56%), *Schilbe intermedius* and *Barbus callipterus* respectively (0.51%), *Distichodus fasciatus* and *Papyrocranys afer* respectively (0.47%), *Ctenopoma petrecii* (0.43%), *Barbus leonensi* (0.34%), *Distichodus rostratus* and *Petrocepahlus microphtalmus* respectively (0.30%), *Hepsetu odoe* and *Xenomystus nigri* (0.25%).

- Four fishing techniques and gear types were used: gillnets, trammel nets, cast nets, and traps. Using gillnets and cast nets, we were able to capture 56.64% and 25.6% of the fish, respectively. These are the two most efficient fishing techniques in this environment.

Regarding metric and meristic characteristics, we focused on four target species (*Chrysichthys wagenaari, Gnathonemus numenius*, *Labeo lineatus,* and *Clarias gariepinus*).

It appears that in *Chrysichthys wagenaari,* males are larger (569.7 mm) than females (501.3 mm) in total length. The sex ratio (M:F) is 1.5:1. The average weight is 571.7 g for males and 556.9 g for females.

In *Gnathonemus* numenius, females are larger (132.85 mm) than males (103.60 mm) in total length. The sex ratio (M–F) is 1.1:1. The average weights are 579.7 g and 504.7 g for males and females, respectively.

In *Labeo lineatus,* females are larger (175.50 mm) than males (166.88 mm) in average total length. The sex ratio (M–F) is 1.7:1. Their average weights are 416.1 g and 371.9 g for females and males, respectively.

In *Clarias gariepinus*, males are larger (179.0 mm) than females (164.84 mm) in total length. The sex ratio (M–F) is 1.2:1. The average weights are 458 g and 400 g, respectively, for males and females.

Analysis of the stomach contents of the various preferred fish species shows that they favor food of animal origin (insects, larvae, fish, scales, and other animal matter), the proportions of which vary by species.

Analysis of the fish marketing process shows that the four preferred species are in high demand among the population. The price per kilogram ranges from 9,700 FC to 13,500 FC. The sale of fish enables the farmer to meet household needs.

In light of these results, we make the following recommendations:

1. To the provincial government

- To consider promoting rational fishing methods among the riverside population and fishermen to avoid disrupting the food chain;
- To fund sustainable aquaculture (fish farming) projects to increase fish stocks.

2. To the central government

- To rehabilitate the fish ponds left behind by the colonizers and thereby strengthen food security;
- To provide provincial administrations with the necessary fishing supplies.

3. To future researchers

- To conduct similar studies, taking into account other parameters not included in this study, in order to enrich the database, which has been lacking in this central part of the Congo River.
- To consider studies on the domestication of preferred species.

Bibliographic References

■ Abbaret, J.J. 1982, "Reproduction and Fertility of Freshwater Fish in Côte d'Ivoire," *Rev. Hydrobiol.* Trop. 5 (4)

■ Adité A. May 1990, *Ecological and Fisheries Study of Several Former Lagoons in Southern Benin*, Technical Report No. 1, Benin, PPL-GTZ

■ Ahouansou, S. 2003, *Study of the Ecology and Fishery Production of Lake Toho in Benin*, Master's Thesis, University of Abomey-Calavi, Cotonou (Benin), 89 pp.

■ Anonymous, 2004, *Monograph of Maniema Province*

■ Baijot, E., Moreau, J., & Bouda, S. 1994, *Hydrobiological and Fish Farming Aspects of Water Reservoirs in the Sudano-Sahelian Zone*, Ede: CTA (ACP/EC), 250 pp.

■ Bigorne R., 1987. The genus Mormyrops (Pisces, Mormyridae) in West Africa. Journal of Tropical Hydrobiology, 20: 145–164.

■ Bigorne R., 1989. The genera Brienomyrus and Isichthys (Pisces, Mormyridae) in West Africa. Journal of Tropical Hydrobiology, 22: 317–338.

■ Bigorne R., 1990. Systematic revision of the genus Pollimyrus (Teleostei, Mormyridae) in West Africa. Journal of Tropical Hydrobiology, 23: 313–327.

■ Bigorne R., 2003. Mormyridae; pp. 155–222. In: PUGY D., LEVEQUE C., and TEUGELS G.G., eds. Fauna of Freshwater and Brackish-Water Fishes of West Africa, Vol. 1. Research Institute for Development, Paris; National Museum of Natural History, Paris; Royal Museum for Central Africa, Tervuren. 457 pp.

■ Blache, J. 1964, *The Fishes of the Chad Basin and the Adjacent Mayo Kebi Basin*, DEA Thesis, ORSTOM

■ Boden G., Teugels G.G., and Hopkins C.D., 1997. A systematic revision of the largescaled Marcusenius with description of a new species from Cameroon (Teleostei; Osteoglossomorpha; Mormyridae). Journal of Natural History., 31:1645-1682.

■ Boulenger G.A. 1915, "Catalogue of the Fresh-water Fishes of Africa" in *British Museum (Natural History)*. The Trustees, London, vol. 3, 526 p.

■ Boulenger G.A. 1916, "Catalogue of the Freshwater Fishes of Africa" in *British Museum (Natural History)*, The Trustees, London, Vol. 4, 392 p.

■ Boulenger, G.A. 1901, *The Fishes of the Congo, published by the Independent State of the Congo*, Brussels

■ Boulenger, G.A. 1909, "Catalogue of the Freshwater Fishes of Africa" in *British Museum (Natural History),* The Trustees, London, Vol. 1, 373 pp.

■ Boulenger, G.A. 1911, "Catalogue of the Fresh-water Fishes of Africa" in *British Museum (Natural History*). The Trustees, London, Vol. 2, 529 pp.

- Bruton, M.N. 1979, "The breeding biology and early development of Clarias garias Piscer: Claridae in Lake Sibaya, South Africa, with a review of breeding" in *Species of the subgenus Clarias (Clarias), Trans.* Zool. Soc., London, 35, pp. 1–45
- Crespi V. 1998, "Preliminary study on the fishery resources of the Niger River" in *The Upper Niger National Park*, Guinea, Fish. Manag. Ecol., 5, pp. 201–208
- Cuimat, R. 1975, *Ecological Diagnosis of Salmonids in Rivers*, FAO, Tech. Doc. CECPI 22, 120 pp.
- Dadebo E. 2000, "Reproductive biology and feeding habits of catfish Clarias gariepinus Burchell (Pisces: Clariidae) in Lake Awassa, Ethiopia," *SINET: Ethiopian Journal of Science*, 23, pp. 231–246
- Daget J., Iltis A. 1965, *Fishes of Côte d'Ivoire (Freshwater and Brackish Waters)*, IFAN-DAKAR Press, 385 pp.
- Daget, J. 1954, *The Fish of the Upper Niger*, Thesis, French Institute for Black Africa, 36, 391 pp.
- Daget, J. 1962, "Relationship between scale size and standard length in *Tilapia galilaea* (Art) of the Middle Niger," *Bull. Inst. Fr, Afr (A),* 24 (2), pp. 486–509
- Daget, J., Gosse, J.P., Teugels G.G., *Thy van den Audenrade (D.F.E.) (eds.),* n.d.,: *Checklist of the Freshwater Fishes of Africa,* Paris, ORSTOM, Volume IV, Brussels, ISNB, Tervuren, MRAC, pp. 482–508
- Dajoz, R. 1996, *Précis d'écologie*, 6th ed., Dunod, Paris, 551 pp.
- De Graaf, G.J., Galemoni, F., & Banzoussi, B. 1995, "The artificial reproduction and fingerling production of the African catfish Clarias gariepinus (Burchell 1822)" in *Protected and unprotected ponds. Aquaculture Research*, 26, pp. 233–242
- De Vos, L. 1995, "A systematic revision of the African Schilbeidae (Teleostei, Siluriformes). With an annotated bibliography," *Ann. Mus. Roy. Afr. Centr.*, 271, 450p.
- Decamps, H. 1967, *Ecology of Caddisflies in the Laure Valley (Haute-Pyrénées)*, Ann. Limnol. 3(3), pp. 399–577
- Decremont, 1978, *Handbook of Water Engineering*, Paris, 8th ed., 1200 pp.
- Depasse, P. 1956, "Ichthyological Monograph of the Eastern Province. Description of the Limnological Environment: Program and Implementation in Fishing and Fish Farming," *Bull. Agr. Congo-Belge*, 7(4), pp. 959–1088
- Ellis, A.E., R.J. Roberts, and Tyler, 1989, "The Anatomy and Physiology of Teleosts," in R.J. Roberts (ed.), *Fish Pathology*, 2nd ed.· Baillière Tindall, W.B. Saunders, London, 467 pp.
- FAO, 2005, *Fishery Record Collections FIGIS Data*, Inland Water Collection. Resources and Aquaculture Service (FIRD; FAO FIGIS ROME)
- FAO, 1999, *Aquaculture Production Statistics 1988–1997*
- Gosse J-P. 1963, "The aquatic environment and the ecology of fish in the Yangambi region," *Ann. Mus.*

r. Afr. Centr., Sci. Zool., 116, pp. 113–271

- Gourène, G., Teugels, G.G., and Thys van Den Audenarde, 1995, *Practical Guide to the Identification of Fish of Lake Ayame (Bia River*), Ivory Coast,

- Harder, W. 2000, *The Mormyridae and other Osteoglossomorpha,* CD-ROM, Vol. 2, New York, Springer-Verlag. *World Biodiversity Database*, CD-ROM Series, UNESCO Publishing

- Harder, W. 1975, *Anatomy of Fishes*. E. Sweizerbart, Verlagsbuchhandlung (Nagele V. Obermiller), Stuttgart, Germany, 612 p.

- Haster, A.D. 1957, "The sense organs: olfactory and gustatory senses of fishes" in M.E. Brown (ed.), *Fish Physiology*, Volume II, Academic Press, London, pp. 187–207.

- Hile, R. 1948, *Standardization of methods of expressing lengths and weights of fish*, Soc. 75,

- http://www.Fishebase.org, 2008

- http://fr.wikipedia.org, 2014

- Hulot, A. 1950, "The diet of fish in Central Africa. Potential interest of these fish for economic aquaculture in the Belgian Congo," *Bull. Agric. Congo Belge*, 41 (1), pp. 145–176

- Humblet A. Ghislain and Van Den Bossche, 1974, *Encyclopedia of Tropical Fish*, 63 p.

- Huvel, M. 1970, *Treatise on Fish Farming,* 4th ed., Brussels, Ch. de Wyngaert, 718 pp.

- Jayaram, (K.C), 1966, "Contribution to the study of the fishes of the family Bragidae 2. A systematic account of the African genus with a new classification of the family", *Bull. Ifan.*, A 28 (3), pp.1064-1139

- Jobling, M. 1995, *Environmental Biology of Fishes. Chapman & Hall* (Publ.), London, United Kingdom, 455 pp.

- Kamdem T.A. and Teugels G.G. 1997. Patterns of microhabitat use among fourteen abundant fishes of the lower Ntem River Basin (Cameroon). Aquat. Living Resour., 10: 289–298.

- Kamdem T.A., 1998. Fish biodiversity of the Ntem basin (Cameroon): Taxonomy, Ecology, and Conservation. Thesis, Catholic University of Leuven, 350 p.

- Kramer B., Van der Bank H., and WINK M., 2004. Hippopotamyrus ansorgii species complex in the upper Zambezi River system with a description of a new species, H. szaboi (Mormyridae). Zool. Scr., 33 (1): 1-18.

- Kestemont, P. 2009, *Fish Nutrition*, Interuniversity Course, D.E.S. ULg-FUNDP, Belgium

- L'évêque, C., Bruton, M.N., Sentonongo, G.W., 1988, *Biology and Ecology of African Freshwater Fish*, Working Paper, ORTOM, Paris, pp. 216–508

- Lairs, S., Mailhot, Y., Higgins, R., Bélanger, D., Berthiaume, L., De la Fontaine, Y., and Martineau, D. 1996, "Jaw Ulcers in Atlantic Tomcod, Microgadus tomcod (Walbaum), from the St. Lawrence River," *Journal of Fish Diseases*, 19, pp. 101–107.

■ Lalèyè, P., Chikau, A., Philipart, G., Teugels, G., and Vandewalle, P. 2004, *Study of the ichthyological diversity of the Ouémé River basin in Benin (West Africa)*, Cybium, (28/4), pp. 329–339

■ Lalèyè, P.H. 1995, *Comparative ecology of two species of Chrysichthys, catfish-like fish (Clarotoidae) from the Lake Nokoué-Porto-Novo lagoon complex in Benin*, Ph.D. thesis, University of Liège (Belgium), 152 p.

■ Lauzane, L., 1988, "The feeding habits of African freshwater fish" in LEVEQUE, C; BRUTON, K. N and SSENTONGO, G. W (eds), *Biology and Ecology of African Freshwater Fish*, Collections of Works and Documents, No. 216, 224 pp.

■ Lavandier, P. 1979, *Ecology of a High-Mountain Pyrenean Torrent: The Estaragne*. Doctoral thesis in Science, U.P.S., Toulouse, 530 pp.

■ Levêque, C. and Paugy, D. 2006, "Geographic Distribution and Affinities of African Freshwater Fish," in *Fish of African Continental Waters: Diversity, Ecology, and Human Use*, IRD, Paris, pp. 59–74.

■ Levêque, C. 1990, "Cyprinidae" in *Fauna of Freshwater and Brackish Water Fish.*

■ Lowe Mc Connell R.H. 1987, "Ecological studies" in *Tropical fish communities*, Cambridge University Press, Tropical Biology series, 382 p

■ Manga, Tchomba, J., Mulungo, S.H., Nyongombe, U.N.F. 2014, "Identification of fish caught in the Congo River and sold in the city of Kindu. The cases of Kindu Central Beach and Basoko Beach," *Shaloom Journal,* University of Bunia, Vol., No. 1, pp.

■ Manga, Tchomba, J., Nyongombe U.N.F.; Okitayela O. 2014, "Determination of selected biological parameters observed in Chrysichthys wagenaari (Pisces, Clarotoidae) captured in the Congo River and sold in the central market of Kindu" *Shaloom Journal,* University of Bunia, Vol., No. 1, pp.

■ Matthes H. 1964, "The fish of Lake Tumba and the Ikela region," *Ann. Mus. R. Afr. Centr. Sc. Zool.,* 126, 204 p.

■ Mbega J.D. 2004, *Fish Biodiversity of the Lower Ogooue Basin (Gabon): Taxonomy, Ecology, Biology, and Conservation*, Ph.D. Thesis, Facultés Universitaires Notre Dame de la Paix, Namur (Belgium), 614 pp.

■ Melanie L.J., Stiassny G.G., Tengels, Hopkins Carl D. 2007, *Freshwater and brackish water fish of Lower Guinea, West Central Africa,* IRD publication, 805 pp.

■ Micha, J. C. 1973, *Study of Fish Populations in the Oubangui River and Attempts to Select and Adapt Certain Species to Fish Farming Ponds*, T.F.T./Nogent-sur-Marne, 110 pp.

■ Moreau, J., Arrignon, J., and Jubb, R.A. 1988, "The introduction of alien species into African and adjacent continental waters," *Freshwater Biology and Ecology*, French Institute for Scientific Research for Development in Cooperation, France, pp. 395–426

■ Moriaty (C) 1983, *"The African Pike Hepsetus odoe,"* *Nigerian Field*, 47 (4), pp. 212–221

■ Mulangu, K. 2007, "Fisheries Value: Physicochemical Analyses of the Waters of the Luilu River," *Ann, ISP. Mbuji-Mayi.*, vol. 15, pp. 67–82

■ Mulangu, K. 2012, *Contribution to the study of the diversity and ecology of fish in the Lubilanji River and its tributaries (Kasaï Oriental/DR Congo),*

■ Mulangu, K. 2012, *Contribution to the study of the diversity and ecology of fish in the Lubilanji River and its tributaries (Western Kasai/DR Congo),* Doctoral Thesis (unpublished), UOM. 197 pp.

■ Musala, L. 1989, *Contribution to the study of the ichthyological fauna of the Kisangani area: Ichthyofauna of the Bitubu River (sub-tributary of the Zaire/left bank)*, Monograph (unpublished), Faculty of Sciences, University of Kisangani, 34 pp.

■ Mutambwe, S. 1984, *Contribution to the study of the ecology of the Luki River (a tributary of the Zaire River): Watershed and fish.* Doctoral dissertation, (unpublished), Sabastier University, Toulouse, 214 pp.

■ Ndjadi, N.J., *Ichthyology of the Yoko and Biaro Rivers and Their Tributaries. Biology and Ecology of Clarias gariepinus. Sauvage 1879,* Master's Thesis (unpublished), Faculty of Science, Department of Ecology and Management of Plant Resources, 91 pp.

■ Nelson, J.S. 1994, *Fishes of the World,* 3rd edition, A Wiley Interscience publication. Wiley & Sons. New York, 600 pp.

■ Nelson, J.S., 2006. Fishes of the World. John Wiley and Sons, Inc. New York. 3rd edition, 601 pp.

■ Niyonkuru, C. 2001, *Study of the Spatio-Temporal Variations in the Ichthyological Fauna of Lake Nokoué in the Republic of Benin*, Master's Thesis, Univ. of Abomey-Calavi, Cotonou (Benin), 125 pp.

■ Norman, J.R. 1931, *A History of Fishes*, Frederick A. Stokes Company Publishers, New York, 463 pp.

■ Norris, S. M. 2002, "A revision of African electric catfishes, Family Malapteruridae (Teleostei, Siluriformes), with the establishment of a new genus and the description of fourteen new species, and an annotated bibliography", *Ann. Mus. Roy. Afr. Centr.*, 289, 155 pp.

■ Nyongombe, U. 1993a, *Determination of the feeding activity of fish in the Kamundele River (a tributary of the Tshopo) in Kisangani*, DES Thesis (unpublished), IFA Yangambi, 35 pp.

■ Nyongombe, U. 2014, *Course on Limnology, Ichthyology, and Fish Farming* (unpublished), Kindu University.

■ Nyongombe, U. 1994b, *Contribution to the Ecological and Biological Study of Fish in the Masendula River in Kisangani*, Doctoral Thesis (unpublished), 175 pp.

■ Pellegrin, J., 1928. "Fish of Chiloango and the Congo collected by the expedition of Dr. H. Schauteden (1920–1922)," *Ann. Mus. R. Congo Belge*, Zoo, (I) 3 (1), 50 pp.

■ Poll, M. 1957, "The Genera of Freshwater Fish of Africa," *Ann. Mus. Congo Belge*, SCI Zooli, 54, pp. 1–191

- Poll, M. and Gosse, J.P. 1995, *Genera of Freshwater Fishes of Africa,* Royal Academy of Belgium, 324 pp.

- Poll, M. and Gosse, J.P. 1963, "Contribution to the systematic study of the ichthyological fauna of the Central Congo," *Ann. Mus. R. Afr. Cont.*, No. 8, Zool., 116, pp. 41–101

- Prado, J. and Dremière, P.-Y. 1988, *Practical Guide for the Fisherman*, FAO. Techn. & Documentation - Lavoisier, Paris. 179 pp.

- Rich, L.M. 1992, "Bagridae" In Lévêque C., Pauguy, D., Teugels, G.G. (eds.), *Fauna of Freshwater and Brackish-Water Fishes of West Africa*, Vol. 2, coll. Tropical Fauna, 28. Tervuren, MRAC, Paris, ORSTOM, pp. 395–431

- Risch, L. 1986, "Bagridae," *In* Daget J., Gosse J.-, and Thys van den Audenaerde D.F.E. (eds.) *Checklist of the Freshwater Fishes of Africa (CLOFFA),* Vol. 2. ORSTOM, Paris, and MRAC, Tervuren, pp. 2–3.

- Rodier J. 1984, *Water Analysis: Natural Waters, Wastewater, Seawater*, Bordas, Paris, 1365 pp.

- Scott and Crossman, 1974, *Freshwater Fishes of Canada. Fisheries Research Board of Canada*, Quebec, 1026 pp.

- Seck, A. 1980, *Catalogue of Artisanal Fishing Gear in Senegal*, COPACE/PACE, Fisheries Technology Service, Fisheries Industries Division, FAO, Rome, 111 pp.

- Staskopf, M.K. (ed) 1993, *Fish Medicine*, WB Saunders & Co., Philadelphia, 882 pp.

- Taverne L., 1971a. Note on the systematics of Mormyriformes catfish. The problem of the genera Gnathonemus Gill, Hippopotamyrus. Revue de Zoologie et de Botanique Africaines., 84: 99–110.

- Taverne L., 1971b. Osteology of the genera Marcusenius Gill, Hippopotamyrus Pappenheim, Cyphomyrus Myers, Pollimyrus Taverne, and Brienomyrus Taverne (Pisces, Mormyriformes). Annals of the Royal Museum for Central Africa, Tervuren, Belgium, 188: 1–143.

- Taverne L., 1972. Osteology of the genera Mormyrus Linné, Mormyrops Müller, Hyperopisus Gill, Isichthys Gill, Myomyrus Boulenger, Stomatorhinus Boulenger, and Gymnarchus Cuvier. General considerations on the systematics of fishes of the order Mormyriformes. Annales Musée Royal de l'Afrique Centrale, Sciences Zoologiques, Tervuren, Belgium, 200: 1-194.

- Teugels G.G. 1986, "Clariidae," in Daget J., J. P. Gosse, and D. F. E. Thys van den Audenaerde. 1986 (eds). *Checklist of the Freshwater Fishes of Africa*, CLOFFA 2, ISNB, (Brussels), MRAC (Tervuren), and ORSTOM (Paris), pp. 66–101

- Teugels G.G. and Adriaens D. 2003, "Taxonomy and phylogeny of Clariidae", in Arratia G., Kapoor B.G., Chardon M., and Diogo R. (eds), *Catfishes, Science Publishers*, Enfield, Vol. I, New Hampshire, pp. 465–487

- Teugels G.G. and Guegan J-F. 1994, "Biological diversity of freshwater fish in Lower Guinea and Central Africa" in Teugels et al. (eds), *Biological diversity of freshwater and brackish water fish in Africa. Geographical Syntheses,* Paradis Symposium, *Ann. Roy. Afr. Centr. Mus., Zool.,* 275p.

- Thys van den Audenaerde (D.F.E.), 1966, "The Tilapia (Pisces cichlidae) of Southern Cameroon and Gabon. A Systematic Study," *Ann. Mus. R. Afr. Centr.*, Series in-8°; Sci. Zoole., 153 pp.
- Toshiaki, 1971. "Chemoreception" in W.S. Hoar and D.J. Randall (eds.), *Fish Physiology*, Vol. V, Academic Press, London, pp. 79–114
- Trewavas E. 1983, *Tilapine fishes of the genera Sarotherodon, Oroechromis and Danakilia*, British Mus. Nat. Hist., 878, 583 p.
- Tshibwabwa S.M. 1997, *Systematics of African species of the genus Labeo (Teleostei, Cyprinidae) in the ichthyological regions of Lower Guinea and the Congo*, Doctoral Dissertation, Facultés Universitaires Notre Dame de la Paix, Namur, 512 p.
- Viera da Silva, J. 1979, *Introduction to Ecological Theory,* Masson, Ecology Series, 14, Paris, 112 pp.
- Wamuini, L.S. 2010, *Ichthyofauna of the Inkisi (Lower Congo/DRC): Diversity and Ecology,* Ph.D. Thesis, ULG, Liège, 351 pp.
- Welcomme, R.L. 1985, *Senior Fishery Resources Officer*, *FAO Fisheries Resources and Environment Division*, Rome, pp. 134–189
- Wembodinga, N. 2014, *Ichthyofauna and Biology of Fish in the Sankuru District (Case of the Lokenye and Loami Rivers*), University of Lodja, 172 pp.
- Yalçin Ş., Solak K., & Akyurt I. 2001, *Certain reproductive characteristics of the catfish (Clarias gariepinus Burchell, 1822) living in the Asi River*, Turkey, Turk J. Zool., 25, pp. 453–460.

APPENDICES

APPENDIX 1: Statistics on the standard length of the species *Clarias gariepinus.*

Table 1: Statistics on the standard length of the species *Clarias gariepinus.*

Standard length (mm)	Number of individuals	Percentage	Cumulative percentage
76	1	1.9	1.9
82	1	1.9	3.8
106	1	1.9	5.7
108	1	1.9	7.5
110	1	1.9	9.4
111	1	1.9	11.3
114	1	1.9	13.2
118	1	1.9	15.1
124	1	1.9	17.0
126	1	1.9	18.9
131	1	1.9	20.8
133	1	1.9	22.6
136	1	1.9	24.5
137	1	1.9	26.4
140	1	1.9	28.3
141	2	3.8	32.1
142	1	1.9	34.0
143	1	1.9	35.8
145	2	3.8	39.6
146	1	1.9	41.5
148	1	1.9	43.4
149	2	3.8	47.2
150	2	3.8	50.9
151	1	1.9	52.8
152	2	3.8	56.6
153	2	3.8	60.4
154	2	3.8	64.2
157	1	1.9	66.0
160	1	1.9	67.9
165	1	1.9	69.8
171	1	1.9	71.7
172	1	1.9	73.6
175	1	1.9	75.5
178	1	1.9	77.4
181	1	1.9	79.2
184	1	1.9	81.1
190	1	1.9	83.0
191	1	1.9	84.9
193	1	1.9	86.8
197	1	1.9	88.7
201	1	1.9	90.6
202	1	1.9	92.5
203	1	1.9	94.3
213	1	1.9	96.2
216	1	1.9	98.1
243	1	1.9	100.0
Total	53	100.0	

Tableau 2 Statistics on the total length of the species *Clarias gariepinus*

Total length (mm)	Number of individuals	Percentage	Cumulative percentage
95	1	1.9	1.9
102	1	1.9	3.8
122	1	1.9	5.7
126	1	1.9	7.5
127	1	1.9	9.4
128	1	1.9	11.3
133	1	1.9	13.2
134	1	1.9	15.1

139	1	1.9	17.0
143	1	1.9	18.9
144	1	1.9	20.8
146	1	1.9	22.6
151	1	1.9	24.5
153	2	3.8	28.3
158	1	1.9	30.2
160	1	1.9	32.1
161	2	3.8	35.8
162	1	1.9	37.7
163	2	3.8	41.5
165	1	1.9	43.4
168	3	5.7	49.1
169	1	1.9	50.9
170	2	3.8	54.7
171	3	5.7	60.4
172	2	3.8	64.2
174	1	1.9	66.0
175	1	1.9	67.9
185	1	1.9	69.8
186	1	1.9	71.7
192	1	1.9	73.6
196	2	3.8	77.4
200	1	1.9	79.2
201	1	1.9	81.1
208	2	3.8	84.9
215	1	1.9	86.8
220	1	1.9	88.7
222	2	3.8	92.5
224	1	1.9	94.3
235	1	1.9	96.2
236	1	1.9	98.1
265	1	1.9	100.0
Total	53	100.0	

Tableau 3 : Weight statistics for the species *Clarias gariepinus*

Weight (g)	Number of individuals	Percentage	Cumulative percentage
9	1	1.9	1.9
11	1	1.9	3.8
12	1	1.9	5.7
14	1	1.9	7.5
15	1	1.9	9.4
17	2	3.8	13.2
18	1	1.9	15.1
21	1	1.9	17.0
25	2	3.8	20.8
26	1	1.9	22.6
28	1	1.9	24.5
29	5	9.4	34.0
30	2	3.8	37.7
31	1	1.9	39.6
32	3	5.7	45.3
33	2	3.8	49.1
34	2	3.8	52.8
35	1	1.9	54.7
39	2	3.8	58.5
41	2	3.8	62.3
42	1	1.9	64.2
44	1	1.9	66.0
46	1	1.9	67.9
48	1	1.9	69.8
52	1	1.9	71.7
53	2	3.8	75.5
56	2	3.8	79.2
59	1	1.9	81.1
65	1	1.9	83.0

66	1	1.9	84.9
73	1	1.9	86.8
78	1	1.9	88.7
79	2	3.8	92.5
91	1	1.9	94.3
98	1	1.9	96.2
117	1	1.9	98.1
126	1	1.9	100.0
Total	53	100.0	

Tableau 4 : Statistics on the standard length of male *Clarias gariepinus.*

Standard length (mm)	Number	Percentage	Cumulative percentage
76	1	3.7	3.7
82	1	3.7	7.4
124	1	3.7	11.1
141	2	7.4	18.5
143	1	3.7	22.2
148	1	3.7	25.9
149	1	3.7	29.6
150	1	3.7	33.3
152	2	7.4	40.7
153	2	7.4	48.1
154	1	3.7	51.9
157	1	3.7	55.6
160	1	3.7	59.3
165	1	3.7	63.0
175	1	3.7	66.7
178	1	3.7	70.4
181	1	3.7	74.1
184	1	3.7	77.8
191	1	3.7	81.5
193	1	3.7	85.2
197	1	3.7	88.9
201	1	3.7	92.6
203	1	3.7	96.3
216	1	3.7	100.0
Total	27	100.0	

Tableau 5 : Statistics on the total length of male *Clarias gariepius*

Total length	Number	Percentage	Cumulative percentage
95	1	3.7	3.7
102.00	1	3.7	7.4
139.00	1	3.7	11.1
160.00	1	3.7	14.8
162.00	1	3.7	18.5
168.00	2	7.4	25.9
169.00	1	3.7	29.6
170.00	2	7.4	37.0
171.00	3	11.1	48.1
172.00	2	7.4	55.6
175.00	1	3.7	59.3
185.00	1	3.7	63.0
192.00	1	3.7	66.7
196.00	1	3.7	70.4
200.00	1	3.7	74.1
201.00	1	3.7	77.8
208.00	1	3.7	81.5
215.00	1	3.7	85.2
220.00	1	3.7	88.9
222.00	1	3.7	92.6
224.00	1	3.7	96.3
235.00	1	3.7	100.0
Total	27	100.0	

Tableau 6 : Statistics on the standard length of female *Clarias gariepinus*

Standard length (mm)	Number of	Percentage	Cumulative percentage

	individuals		
106.00	1	3.8	3.8
108.00	1	3.8	7.7
110.00	1	3.8	11.5
111.00	1	3.8	15.4
114.00	1	3.8	19.2
118.00	1	3.8	23.1
126.00	1	3.8	26.9
131.00	1	3.8	30.8
133.00	1	3.8	34.6
136.00	1	3.8	38.5
137.00	1	3.8	42.3
140.00	1	3.8	46.2
142.00	1	3.8	50.0
145.00	2	7.7	57.7
146.00	1	3.8	61.5
149.00	1	3.8	65.4
150.00	1	3.8	69.2
151.00	1	3.8	73.1
154.00	1	3.8	76.9
171.00	1	3.8	80.8
172.00	1	3.8	84.6
190.00	1	3.8	88.5
202.00	1	3.8	92.3
213.00	1	3.8	96.2
243.00	1	3.8	100.0
Total	26	100.0	

Tableau 7 Statistics on the total length of the species

Female *Clarias gariepinus*

Total length (mm)	Number of individuals	Percentage	Cumulative percentage
122.00	1	3.8	3.8
126.00	1	3.8	7.7
127.00	1	3.8	11.5
128.00	1	3.8	15.4
133.00	1	3.8	19.2
134.00	1	3.8	23.1
143.00	1	3.8	26.9
144.00	1	3.8	30.8
146.00	1	3.8	34.6
151.00	1	3.8	38.5
153.00	2	7.7	46.2
158.00	1	3.8	50.0
161.00	2	7.7	57.7
163.00	2	7.7	65.4
$165.00	1	3.8	69.2
168.00	1	3.8	73.1
174.00	1	3.8	76.9
186.00	1	3.8	80.8
196.00	1	3.8	84.6
208.00	1	3.8	88.5
222.00	1	3.8	92.3
236.00	1	3.8	96.2
265.00	1	3.8	100.0
Total	26	100.0	

Tableau 8 : Weight statistics for female *Clarias gariepinus*

Weight (g)	Number	Percentage	Cumulative percentage
9	1	3.8	3.8
11	1	3.8	7.7
14	1	3.8	11.5
15	1	3.8	15.4
18	1	3.8	19.2
21	1	3.8	23.1

25	1	3.8	26.9
26	1	3.8	30.8
28	1	3.8	34.6
29	2	7.7	42.3
30	1	3.8	46.2
31	1	3.8	50.0
32	3	11.5	61.5
35	1	3.8	65.4
39	1	3.8	69.2
41	1	3.8	73.1
44	1	3.8	76.9
46	1	3.8	80.8
48	1	3.8	84.6
65	1	3.8	88.5
98	1	3.8	92.3
117	1	3.8	96.2
126	1	3.8	100.0
Total	26	100.0	

Tableau 9 : Weight statistics for male *Clarias gariepinus*

Weight	Number	Percentage	Cumulative percentage
12.00	1	3.7	3.7
17.00	2	7.4	11.1
25.00	1	3.7	14.8
29.00	3	11.1	25.9
30.00	1	3.7	29.6
33.00	2	7.4	37.0
34.00	2	7.4	44.4
39.00	1	3.7	48.1
41.00	1	3.7	51.9
42.00	1	3.7	55.6
52.00	1	3.7	59.3
53.00	2	7.4	66.7
56.00	2	7.4	74.1
59.00	1	3.7	77.8
66.00	1	3.7	81.5
73.00	1	3.7	85.2
78.00	1	3.7	88.9
79.00	2	7.4	96.3
91.00	1	3.7	100.0
Total	27	100.0	

Tableau 10 : Overall statistics on the standard length of the species *Labeo lineatus*

Standard length (mm)	Number of individuals	Percentage	Cumulative percentage
96	1	2.0	2.0
99	1	2.0	4.0
103	1	2.0	6.0
106	1	2.0	8.0
108	1	2.0	10.0
113	1	2.0	12.0
116	1	2.0	14.0
120	2	4.0	18.0
121	1	2.0	20.0
125	1	2.0	22.0
130	1	2.0	24.0
131	2	4.0	28.0
134	1	2.0	30.0
136	1	2.0	32.0
137	1	2.0	34.0
142	2	4.0	38.0
144	2	4.0	42.0
147	1	2.0	44.0
148	1	2.0	46.0
149	3	6.0	52.0
150	2	4.0	56.0

151	2	4.0	60.0
153	1	2.0	62.0
155	1	2.0	64.0
156	1	2.0	66.0
157	1	2.0	68.0
161	2	4.0	72.0
165	2	4.0	76.0
166	1	2.0	78.0
169	1	2.0	80.0
171	1	2.0	82.0
178	1	2.0	84.0
187	1	2.0	86.0
199	1	2.0	88.0
210	2	4.0	92.0
219	1	2.0	94.0
226	1	2.0	96.0
227	1	2.0	98.0
228	1	2.0	100.0
Total	50	100.0	

Tableau 11 : Overall statistics on the total length of the species *Labeo lineatus*

Total length (mm)	Number of individuals	Percentage	Cumulative percentage
105	1	2.0	2.0
117	1	2.0	4.0
118	1	2.0	6.0
121	1	2.0	8.0
126	1	2.0	10.0
133	1	2.0	12.0
134	1	2.0	14.0
135	1	2.0	16.0
136	1	2.0	18.0
137	1	2.0	20.0
141	1	2.0	22.0
142	1	2.0	24.0
146	1	2.0	26.0
147	1	2.0	28.0
149	1	2.0	30.0
151	1	2.0	32.0
154	1	2.0	34.0
157	1	2.0	36.0
160	1	2.0	38.0
161	1	2.0	40.0
162	1	2.0	42.0
163	1	2.0	44.0
167	2	4.0	48.0
168	1	2.0	50.0
169	3	6.0	56.0
170	1	2.0	58.0
171	2	4.0	62.0
172	1	2.0	64.0
175	1	2.0	66.0
177	2	4.0	70.0
178	1	2.0	72.0
182	1	2.0	74.0
186	2	4.0	78.0
188	1	2.0	80.0
190	1	2.0	82.0
201	1	2.0	84.0
207	1	2.0	86.0
218	1	2.0	88.0
233	2	4.0	92.0
240	1	2.0	94.0
241	1	2.0	96.0
246	1	2.0	98.0
253	1	2.0	100.0

Total	50	100.0	

Tableau 12 : Overall statistics on the weight of the species *Labeo lineatus*

Weight	Number	Percentage	Cumulative percentage
8	1	2.0	2.0
9	1	2.0	4.0
11	1	2.0	6.0
14	1	2.0	8.0
15	1	2.0	10.0
18	2	4.0	14.0
19	1	2.0	16.0
21	2	4.0	20.0
23	1	2.0	22.0
25	2	4.0	26.0
26	1	2.0	28.0
28	1	2.0	30.0
29	3	6.0	36.0
30	3	6.0	42.0
31	2	4.0	46.0
32	5	10.0	56.0
33	1	2.0	58.0
34	1	2.0	60.0
35	2	4.0	64.0
37	2	4.0	68.0
38	1	2.0	70.0
39	1	2.0	72.0
41	2	4.0	76.0
44	1	2.0	78.0
46	2	4.0	82.0
48	1	2.0	84.0
65	1	2.0	86.0
66	1	2.0	88.0
70	1	2.0	90.0
74	1	2.0	92.0
89	1	2.0	94.0
98	1	2.0	96.0
117	1	2.0	98.0
126	1	2.0	100.0
Total	50	100.0	

Tableau 13 : Statistics on the standard length of the male *Labeo lineatus*

Standard length	Number	Percentage	Cumulative percentage
96	1	3.1	3.1
103	1	3.1	6.3
106	1	3.1	9.4
113	1	3.1	12.5
120	1	3.1	15.6
121	1	3.1	18.8
125	1	3.1	21.9
130	1	3.1	25.0
131	1	3.1	28.1
134	1	3.1	31.3
136	1	3.1	34.4
137	1	3.1	37.5
142	2	6.3	43.8
144	2	6.3	50.0
147	1	3.1	53.1
148	1	3.1	56.3
149	1	3.1	59.4
150	2	6.3	65.6
151	1	3.1	68.8
155	1	3.1	71.9
156	1	3.1	75.0
165	1	3.1	78.1
166	1	3.1	81.3
171	1	3.1	84.4

178	1	3.1	87.5
187	1	3.1	90.6
226	1	3.1	93.8
227	1	3.1	96.9
228	1	3.1	100.0
Total	32	100.0	

Tableau 14 : Statistics on the total length of the species *Labeo lineatus* male

Total length	Number	Percentage	Cumulative percentage
105	1	3.1	3.1
117	1	3.1	6.3
121	1	3.1	9.4
133	1	3.1	12.5
135	1	3.1	15.6
136	1	3.1	18.8
141	1	3.1	21.9
142	1	3.1	25.0
146	1	3.1	28.1
149	1	3.1	31.3
151	1	3.1	34.4
154	1	3.1	37.5
157	1	3.1	40.6
160	1	3.1	43.8
161	1	3.1	46.9
162	1	3.1	50.0
163	1	3.1	53.1
167	1	3.1	56.3
168	1	3.1	59.4
170	1	3.1	62.5
171	1	3.1	65.6
172	1	3.1	68.8
175	1	3.1	71.9
177	1	3.1	75.0
182	1	3.1	78.1
188	1	3.1	81.3
190	1	3.1	84.4
201	1	3.1	87.5
207	1	3.1	90.6
240	1	3.1	93.8
246	1	3.1	96.9
253	1	3.1	100.0
Total	32	100.0	

Tableau 15 : Weight statistics for the species *Labeo* lineatus (male)

Weight	Number	Percentage	Cumulative percentage
8	1	3.1	3.1
9	1	3.1	6.3
11	1	3.1	9.4
14	1	3.1	12.5
15	1	3.1	15.6
18	2	6.3	21.9
21	2	6.3	28.1
23	1	3.1	31.3
25	1	3.1	34.4
26	1	3.1	37.5
28	1	3.1	40.6
29	2	6.3	46.9
30	1	3.1	50.0
31	1	3.1	53.1
32	3	9.4	62.5
33	1	3.1	65.6
35	1	3.1	68.8
39	1	3.1	71.9
41	1	3.1	75.0
44	1	3.1	78.1
46	2	6.3	84.4

48	1	3.1	87.5
65	1	3.1	90.6
98	1	3.1	93.8
117	1	3.1	96.9
126	1	3.1	100.0
Total	32	100.0	

Tableau 16 Overall statistics on the standard length of the female *Labeo* lineatus

Standard length	Number	Percentage	Cumulative percentage
99	1	5.6	5.6
108	1	5.6	11.1
116	1	5.6	16.7
120	1	5.6	22.2
131	1	5.6	27.8
149	2	11.1	38.9
151	1	5.6	44.4
153	1	5.6	50.0
157	1	5.6	55.6
161	2	11.1	66.7
165	1	5.6	72.2
169	1	5.6	77.8
199	1	5.6	83.3
210	2	11.1	94.4
219	1	5.6	100.0
Total	18	100.0	

Tableau 17 : Overall statistics on the total length of the female *Labeo lineatus*

Total length	Number	Percentage	Cumulative percentage
118	1	5.6	5.6
126	1	5.6	11.1
134	1	5.6	16.7
137	1	5.6	22.2
147	1	5.6	27.8
167	1	5.6	33.3
169	3	16.7	50.0
171	1	5.6	55.6
177	1	5.6	61.1
178	1	5.6	66.7
186	2	11.1	77.8
218	1	5.6	83.3
233	2	11.1	94.4
241	1	5.6	100.0
Total	18	100.0	

Table 18: Overall statistics on the weight of female *Labeo lineatus*

Weight (g)	Number	Percentage	Cumulative percentage
19	1	5.6	5.6
25	1	5.6	11.1
29	1	5.6	16.7
30	2	11.1	27.8
31	1	5.6	33.3
32	2	11.1	44.4
34	1	5.6	50.0
35	1	5.6	55.6
37	2	11.1	66.7
38	1	5.6	72.2
41	1	5.6	77.8
66	1	5.6	83.3
70	1	5.6	88.9
74	1	5.6	94.4
89	1	5.6	100.0
Total	18	100.0	

Table 19: Overall statistics on the standard length of the species *Gnathonemus numenius*

standard length	Number	Percentage	Cumulative percentage

11	1	4	4
75	15	5.3	5.7
80	44	15.7	21.4
85	15	5.3	26.7
92	16	5.7	32.4
95	47	16.7	49.1
100	41	14.6	63.7
101	26	9.3	73.0
105	8	2.8	75.8
109	10	3.6	79.4
111	18	6.4	85.8
115	13	4.6	90.4
132	14	5.0	95.4
142	13	4.6	100.0
Total	281	100.0	

Tableau 20 : Overall statistics on the total length of the species *Gnathonemus numenius*

Total length	Population	Percentage	Cumulative percentage
91.00	17	6.0	6.0
95.00	44	15.7	21.7
100.00	14	5.0	26.7
108.00	16	5.7	32.4
111.00	31	11.0	43.4
114.00	15	5.3	48.8
117.00	14	5.0	53.7
120.00	8	2.8	56.6
121.00	14	5.0	61.6
125.00	30	10.7	72.2
126.00	10	3.6	75.8
127.00	14	5.0	80.8
129.00	13	4.6	85.4
138.00	15	5.3	90.7
152.00	13	4.6	95.4
162.00	13	4.6	100.0
Total	281	100.0	

Tableau 21 : Overall statistics on the weight of the species *Gnathonemus numenius*

Weight	Number of individuals	Percentage	Cumulative percentage
8	1	4	4
9	2	7	1.1
17	2	7	1.8
18	2	7	2.5
21	3	1.1	3.6
22	1	4	3.9
23	3	1.1	5.0
25	5	1.8	6.8
28	4	1.4	8.2
29	10	3.6	11.7
30	3	1.1	12.8
31	4	1.4	14.2
32	6	2.1	16.4
33	11	3.9	20.3
34	10	3.6	23.8
35	7	2.5	26.3
37	6	2.1	28.5
38	2	7	29.2
39	4	1.4	30.6
41	6	2.1	32.7
42	1	4	33.1
43	9	3.2	36.3
44	13	4.6	40.9
45	9	3.2	44.1
46	5	1.8	45.9
47	1	4	46.3
48	1	4	46.6
49	2	7	47.3

50	1	4	47.7
51	2	7	48.4
52	1	4	48.8
53	3	1.1	49.8
54	6	2.1	52.0
55	11	3.9	55.9
56	12	4.3	60.1
57	2	7	60.9
58	1	4	61.2
59	3	1.1	62.3
61	1	4	62.6
65	10	3.6	66.2
66	10	3.6	69.8
67	10	3.6	73.3
68	1	4	73.7
70	3	1.1	74.7
72	1	4	75.1
73	1	4	75.4
74	2	7	76.2
76	8	2.8	79.0
77	10	3.6	82.6
78	10	3.6	86.1
79	2	7	86.8
80	1	4	87.2
85	5	1.8	89.0
86	1	4	89.3
87	7	2.5	91.8
88	4	1.4	93.2
89	5	1.8	95.0
90	1	4	95.4
91	1	4	95.7
95	1	4	96.1
97	1	4	96.4
98	5	1.8	98.2
99	2	7	98.9
117	2	7	99.6
121	1	4	100.0
Total	281	100.0	

Tableau 22 : Overall statistics on the weight of the species *Gnathonemus numenius*

Weight	Number	Percentage	Cumulative percentage
9	1	7	7
17	2	1.3	2.0
18	1	7	2.6
21	1	7	3.3
23	1	7	4.0
25	3	2.0	6.0
28	2	1.3	7.3
29	6	4.0	11.3
30	2	1.3	12.6
31	2	1.3	13.9
32	3	2.0	15.9
33	4	2.6	18.5
34	6	4.0	22.5
35	3	2.0	24.5
37	3	2.0	26.5
38	1	.7	27.2
39	1	.7	27.8
41	3	2.0	29.8
42	1	.7	30.5
43	2	1.3	31.8
44	5	3.3	35.1
45	3	2.0	37.1
46	2	1.3	38.4
47	1	.7	39.1
48	1	.7	39.7

49	1	.7	40.4
51	1	.7	41.1
52	1	.7	41.7
53	3	2.0	43.7
54	2	1.3	45.0
55	4	2.6	47.7
56	5	3.3	51.0
57	2	1.3	52.3
59	3	2.0	54.3
65	8	5.3	59.6
66	7	4.6	64.2
67	7	4.6	68.9
68	1	.7	69.5
70	1	.7	70.2
73	1	.7	70.9
74	1	.7	71.5
76	3	2.0	73.5
77	4	2.6	76.2
78	5	3.3	79.5
79	2	1.3	80.8
80	1	.7	81.5
85	5	3.3	84.8
87	4	2.6	87.4
88	4	2.6	90.1
89	3	2.0	92.1
90	1	.7	92.7
91	1	.7	93.4
95	1	.7	94.0
97	1	.7	94.7
98	4	2.6	97.4
99	2	1.3	98.7
117	1	.7	99.3
121	1	.7	100.0
Total	151	100.0	

Tableau 23 : Overall statistics on the weight of female *Gnathonemus numenius*

Weight	Number	Percentage	Cumulative percentage
8	1	0.8	8
9	1	0.8	1.5
18	1	0.8	2.3
21	2	1.5	3.8
22	1	0.8	4.6
23	2	1.5	6.2
25	2	1.5	7.7
28	2	1.5	9.2
29	4	3.1	12.3
30	1	0.8	13.1
31	2	1.5	14.6
32	3	2.3	16.9
33	7	5.4	22.3
34	4	3.1	25.4
35	4	3.1	28.5
37	3	2.3	30.8
38	1	.8	31.5
39	3	2.3	33.8
41	3	2.3	36.2
43	7	5.4	41.5
44	8	6.2	47.7
45	6	4.6	52.3
46	3	2.3	54.6
49	1	.8	55.4
50	1	.8	56.2
51	1	.8	56.9
54	4	3.1	60.0
55	7	5.4	65.4
56	7	5.4	70.8

58	1	.8	71.5
61	1	.8	72.3
65	2	1.5	73.8
66	3	2.3	76.2
67	3	2.3	78.5
70	2	1.5	80.0
72	1	.8	80.8
74	1	.8	81.5
76	5	3.8	85.4
77	6	4.6	90.0
78	5	3.8	93.8
86	1	.8	94.6
87	3	2.3	96.9
89	2	1.5	98.5
98	1	.8	99.2
117	1	.8	100.0
Total	130	100.0	

Tableau 24 : Overall statistics on the standard length of the species *Chrysichtys wagenaari.*

Standard length	Number	Percentage	Cumulative percentage
32	1	1.7	1.7
33	1	1.7	3.3
35	4	6.7	10.0
37	1	1.7	11.7
38	3	5.0	16.7
39	6	10.0	26.7
40	4	6.7	33.3
41	3	5.0	38.3
42	4	6.7	45.0
43	6	10.0	55.0
44	5	8.3	63.3
45	4	6.7	70.0
47	1	1.7	71.7
48	2	3.3	75.0
49	1	1.7	76.7
52	3	5.0	81.7
53	4	6.7	88.3
55	2	3.3	91.7
58	2	3.3	95.0
60	2	3.3	98.3
65	1	1.7	100.0
Total	60	100.0	

Tableau 25 : Overall statistics on the total length of the species *Chrysichtys wagenaari.*

Total length	Number	Percentage	Cumulative percentage
42	3	5.0	5.0
43	1	1.7	6.7
45	1	1.7	8.3
46	7	11.7	20.0
47	4	6.7	26.7
48	5	8.3	35.0
49	2	3.3	38.3
50	1	1.7	40.0
54	4	6.7	46.7
55	6	10.0	56.7
56	2	3.3	60.0
57	2	3.3	63.3
58	3	5.0	68.3
59	4	6.7	75.0
60	3	5.0	80.0
61	2	3.3	83.3
62	3	5.0	88.3
65	2	3.3	91.7
67	2	3.3	95.0
68	2	3.3	98.3
72	1	1.7	100.0

Total	60	100.0	

Tableau 26 : Overall statistics on the weight of the species *Chrysichtys wagenaari.*

Weight	Number of individuals	Percentage	Cumulative percentage
34	3	5.0	5.0
37	1	1.7	6.7
42	2	3.3	10.0
43	3	5.0	15.0
44	3	5.0	20.0
45	4	6.7	26.7
47	2	3.3	30.0
48	2	3.3	33.3
50	1	1.7	35.0
53	1	1.7	36.7
54	8	13.3	50.0
55	4	6.7	56.7
56	7	11.7	68.3
64	1	1.7	70.0
65	6	10.0	80.0
66	1	1.7	81.7
67	1	1.7	83.3
68	1	1.7	85.0
70	1	1.7	86.7
76	2	3.3	90.0
77	1	1.7	91.7
78	2	3.3	95.0
80	1	1.7	96.7
88	1	1.7	98.3
98	1	1.7	100.0
Total	60	100.0	

Tableau 27 : Overall statistics on the standard length of the male *Chrysichtys wagenaari*

Standard length	Headcount	Percentage	Cumulative percentage
32	1	1.7	1.7
33	1	1.7	3.3
35	4	6.7	10.0
37	1	1.7	11.7
38	3	5.0	16.7
39	6	10.0	26.7
40	4	6.7	33.3
41	3	5.0	38.3
42	4	6.7	45.0
43	6	10.0	55.0
44	5	8.3	63.3
45	4	6.7	70.0
47	1	1.7	71.7
48	2	3.3	75.0
49	1	1.7	76.7
52	3	5.0	81.7
53	4	6.7	88.3
55	2	3.3	91.7
58	2	3.3	95.0
60	2	3.3	98.3
65	1	1.7	100.0
Total	60	100.0	

Tableau 28 : Overall statistics on the total length of the male *Chrysichtys wagenaari*

Total length	Number	Percentage	Cumulative percentage
42	1	2.8	2.8
45	1	2.8	5.6
46	2	5.6	11.1
47	2	5.6	16.7
48	1	2.8	19.4
54	2	5.6	25.0
55	5	13.9	38.9
56	1	2.8	41.7

57	1	2.8	44.4
58	3	8.3	52.8
59	4	11.1	63.9
60	3	8.3	72.2
61	2	5.6	77.8
62	3	8.3	86.1
65	2	5.6	91.7
67	2	5.6	97.2
72	1	2.8	100.0
Total	36	100.0	

Tableau 29 : Overall statistics on the weight of the male *Chrysichtys wagenaari*

Weight	Population	Percentage	Cumulative percentage
34	2	5.6	5.6
37	1	2.8	8.3
42	2	5.6	13.9
43	2	5.6	19.4
44	1	2.8	22.2
45	3	8.3	30.6
47	1	2.8	33.3
48	1	2.8	36.1
50	1	2.8	38.9
53	1	2.8	41.7
54	4	11.1	52.8
55	1	2.8	55.6
56	4	11.1	66.7
64	1	2.8	69.4
65	3	8.3	77.8
66	1	2.8	80.6
67	1	2.8	83.3
70	1	2.8	86.1
76	2	5.6	91.7
78	2	5.6	97.2
88	1	2.8	100.0
Total	36	100.0	

Tableau 30 : Overall statistics on the standard length of the female *Chrysichtys wagenaari*

standard length	Number	Percentage	Cumulative percentage
32	1	4.2	4.2
35	3	12.5	16.7
38	3	12.5	29.2
39	3	12.5	41.7
40	3	12.5	54.2
41	3	12.5	66.7
43	3	12.5	79.2
44	1	4.2	83.3
45	2	8.3	91.7
58	2	8.3	100.0
Total	24	100.0	

Tableau 31 : Overall statistics on the total length of the female *Chrysichtys wagenaari*

Total length	Number	Percentage	Cumulative percentage
42	2	8.3	8.3
43	1	4.2	12.5
46	5	20.8	33.3
47	2	8.3	41.7
48	4	16.7	58.3
49	2	8.3	66.7
50	1	4.2	70.8
54	2	8.3	79.2
55	1	4.2	83.3
56	1	4.2	87.5
57	1	4.2	91.7
68	2	8.3	100.0
Total	24	100.0	

Tableau 32 Overall statistics on the weight of the female *Chrysichtys wagenaari*

weight	Population	Percentage	Cumulative percentage
34	1	4.2	4.2
43	1	4.2	8.3
44	2	8.3	16.7
45	1	4.2	20.8
47	1	4.2	25.0
48	1	4.2	29.2
54	4	16.7	45.8
55	3	12.5	58.3
56	3	12.5	70.8
65	3	12.5	83.3
68	1	4.2	87.5
77	1	4.2	91.7
80	1	4.2	95.8
98	1	4.2	100.0
Total	24	100.0	

Appendix 2: Different families and their percentages

1. Families	Number	Percentage
Cichlidae	6	17.14%
Clariidae	4	11.43%
Mormyridae	4	11.43%
Clarotoidae	3	8.57%
Cyprin*idae*	3	8.57%
Distichodontidae	2	5.71%
Notopteridae	2	5.71%
Schilbeidae	2	5.71%
Malapteruridae	2	5.71%
Citharinidae	2	5.71%
Hepsetidae	1	2.86%
Anabantidae,	1	2.86%
Arapaimidae	1	2.86%
Mochokidae	1	2.86%
Channid ae.	1	2.86%
Total	35	100%

2. SPECIES	**Number**	**Percentage/Total number of fish**
Clarias gariepinus	200	**8.00%**
Clarias angolesis	140	5.60%
Clarias bithupogan	152	6.08%
Clarias cameruneusis	19	0.76%
Chrysicthys nigrodigitatus	32	1.28%
Chrysicthys auratus	42	1.68%
Chrysicthys wagenaari	60	2.40%
Oreochromis macrochure	49	1.96%
Oreochromis mosambicus	52	2.08%
Tilapia zillii	142	5.68%
Oreochromis niloticus	152	6.08%
Sarotherodom melnotherom	100	4.00%
Sarotherodom galilaeus	110	4.40%
Gnathonemus numenius	280	11.20%
Petrocepahlus ballayi	20	0.80%
Petrocepahlus microphtalmus	7	0.28%
Mormyrops nigrican	13	0.52%
Disticodus rostratus	7	0.28%
Disticodus fasciatus	11	0.44%
Hepsetus odoe	6	0.24%
Ctenopoma petreci	10	0.40%
Xenomystus nigri	6	0.24%
Papyrocranys afer	11	0.44%
Barbus callipterus	12	0.48%
Barbus leonensis	8	0.32%

Labeo lineatus	118	4.72%		
Heterotus nilocutus	222	8.88%		
Schilbe intermedius	12	0.48%		
Schilbe grefeli	15	0.60%		
Electric ray	68	2.72%		
Malpterirus lacepede	42	1.68%		
Citharinus gibbosus	106	4.24%		
Citharinus congicus	62	2.48%		
Synodontis batesii	32	1.28%		
	2,318	100.00%		

3. SPECIES	**Number**	Percentage of family
CLARIAS		
Clarias gariepinus	200	39.14%
Clarias angolensis	140	27.40%
Clarias bithupogon	152	29.75%
Clarias camerunensis	19	3.72%
Subtotal	**511**	100.00%
CLAROTOIDAE		
Chrysicthys nigrodigitatus	32	23.88%
Chrysicthys auratus	42	31.34%
Chrysicthys wagenaari	60	44.78%
Subtotal	**134**	100.00%
CICHLIDAE		
Oreochromis macrochir	49	8.10%
Oreochromis mosambicus	52	8.60%
Tilapia zillii	142	23.47%
Oreochromis niloticus	152	25.12%
Sarotherodom melnotherom	100	16.53%
Sarotherodom galilaeus	110	18.18%
Subtotal	**605**	100.00%
MORMYRIDAE		
Gnathonemus numenius	280	87.50%
Petrocephalus ballayi	20	6.25%
Petrocephalus microphtalmus	7	2.19%
Mormyrops nigrican	13	4.06%
Subtotal	**320**	100.00%
DISTICHONDONTIDAE		
Distichodus rostratus	7	38.89%
Distichodus fasciatus	11	61.11%
Subtotal	**18**	100.00%
HEPSETIDAE		
Hepsetus odoe	6	100.00%
Subtotal	**6**	100.00%
ANABANTIDAE		
Ctenopoma petrerici	10	100.00%
Subtotal	**10**	100.00%
NOTOPTERIDAE		
Xenomystus nigri	6	35.29%
Papyrocranys afer	11	64.71%
Subtotal	**17**	100.00%
CYPRINIDAE		
Barbus callipterus	12	8.70%
Barbus leonensis	8	5.80%
Labeo lineatus	118	85.51%
Subtotal	**138**	100.00%
ARAPAMIDAE		
Heterotus niloticus	222	222.00%
Subtotal	**222**	100.00%
SCHILBEIDAE		
Schilbe intermedius	12	44.44%
Schilbe grefeli	15	55.56%
Subtotal	**27**	100.00%

MALAPTERURIIDAE		
Malpterirus electricus	68	61.82%
Malpterirus lacepede	42	38.18%
Subtotal	**110**	100.00%
CITHARINIDAE		
Citharinus gibbosus	106	63.10%
Citharinus congicus	62	36.90%
Subtotal	**168**	100.00%
MOCHOKIDAE		
Synodontis batesii	32	100.00%
Subtotal	**32**	100.00%
CHANNIDAE		
Parachana obscura	182	100.00%
Subtotal	**182**	100.00%
Grand Total	**2,318**	100.00%

APPENDIX 3: Preferred Fish Species

Chrisychtys wgenaari

Labeo weeksii

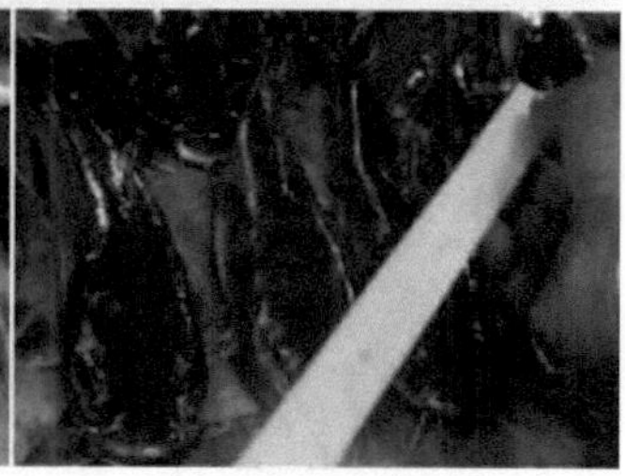

Gnathonemus numenius

Clarias gariepinus

Chrysicthys auratus

Chrysicthys ornatus

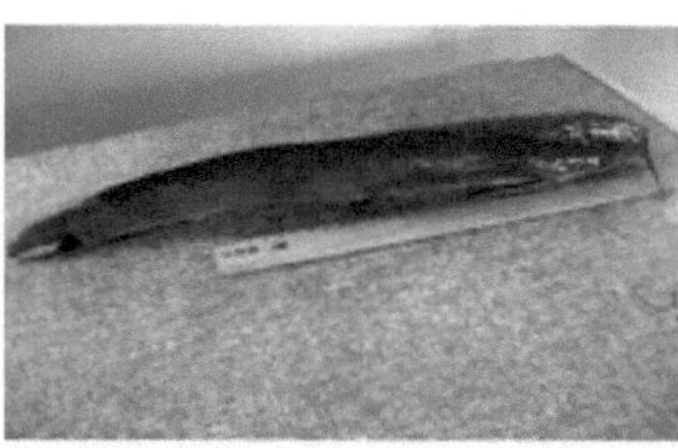

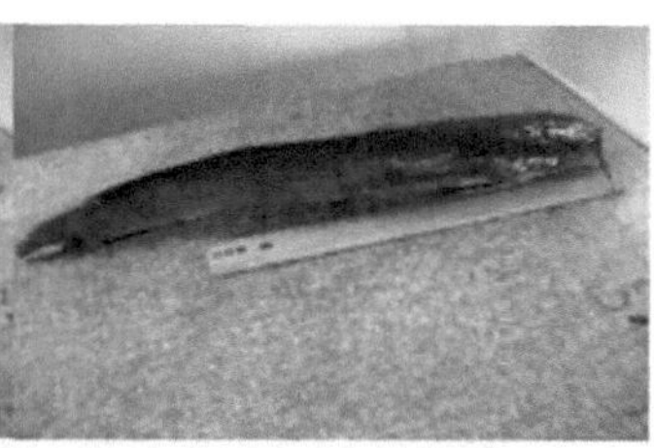

Clarias gariepinus

Clarias angolensis

Malapterus electricus

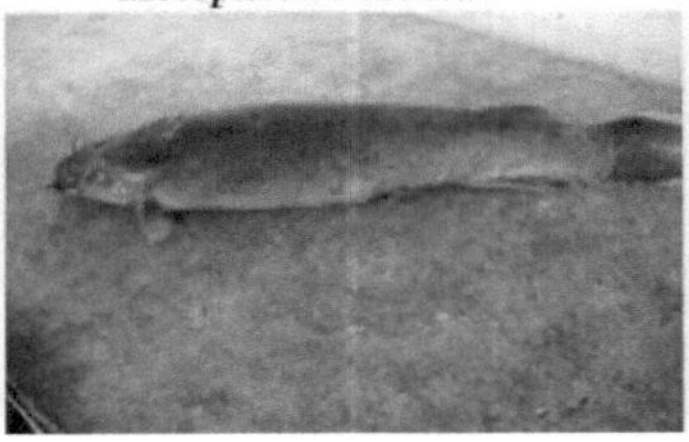

Oreochromis macrochir

Tilapia nilotica

Tilapia rendalli

Citharinus congicus

Labeo lineatus

APPENDIX 4: Survey Questionnaire on Kindu Fishing Activities

1. Since what year have you been fishing?

2. Are there any regulations regarding fishing?

3. Where do you fish?

4. What type of gear do you use? And which is the most efficient?

5. What species of fish do you catch most often?

6. What is the average size of the main fish caught?

7. Apart from the gear-based fishing methods mentioned above, do you use any other fishing techniques? Yes or No. If yes, which one?

8. Do you have an estimate of your daily, monthly, or annual catch? If yes, what is it, and if not, why?

9. At what time of year do you observe fish spawning? What fish are found in Kindu?

10. Do you make a living solely from fishing? Yes or No? If no, what other activity(ies) do you engage in?

11. Are you satisfied with the fishing activity in your family?

12. What challenges do you face in your work as a fisherman?

13. What do you produce?

14. How do you allocate the money earned from sales?

Survey Questionnaire on the Sale (Marketing) of Fish in Kindu

1. Since what year have you been selling fish?

2. What species of fish do you sell?

3. Between naked-bodied fish and scaled fish, which sells better? And why?

5. What hygiene measures do you take to ensure the fish you sell remain clean?

6. Are you satisfied with your family's fish business?

7. Do you make a living solely from the fish business? Yes or no? If not, what other activities do you engage in?

8. What difficulties do you face in your work as a fish seller?

9. Where do the fish sold at the market come from?

10. How do you allocate the money earned from sales?

11. How do you store the rest of your fish?

APPENDIX 5: Summary of Metric and Morphological Data

1. Species: *Clarias gariepinus*

Characteristics observed	Min - Max	Number of fish observed	Mean	Standard deviation	Coeff of variation (%)
Total range	95 - 265	53	172.1	34.69	20.15
Long stand	76–243	53	154	33.51	21.75

2. Species *Labeo lineatus*

Observed characteristics	Min - Max	Number of fish observed	Mean	Standard deviation	Coeff of variation (%)
Total length	105 - 253	50	169.98	35.35	20.79
Long stand	96 - 228	50	152.10	33.54	22.05

3. Species *Gnathonemus numenius*

Observed characteristics	Min - Max	Number of fish observed	Mean	Standard deviation	Coeff of variation (%)
Total length	91–162	281	117.13	18.37	15.68
Long stand	11 - 142	281	98.62	17.25	17.49

4. Species *Chrysichtys wagenaari*

Characteristics observed	Min - Max	Number of fish observed	Average	Standard deviation	Coeff of variation (%)
Total range	42–72	60	54.23	7.58	13.97
Long stand	32–65	60	44.50	7.32	16.44

MIX
Papier aus verantwortungsvollen Quellen
Paper from responsible sources
FSC® C105338

Printed by Books on Demand GmbH, Norderstedt / Germany